T0229801

Artificial Intelligence, Blockchain and IoT for Smart Healthcare

RIVER PUBLISHERS SERIES IN INFORMATION SCIENCE AND TECHNOLOGY

Series Editors:

K. C. Chen
National Taiwan University, Taipei, Taiwan
and
University of South Florida, USA

Sandeep Shukla
Virginia Tech, USA
and
Indian Institute of Technology Kanpur, India

Indexing: All books published in this series are submitted to the Web of Science Book Citation Index (BkCI), to SCOPUS, to CrossRef and to Google Scholar for evaluation and indexing.

The "River Publishers Series in Information Science and Technology" covers research which ushers the 21st Century into an Internet and multimedia era. Multimedia means the theory and application of filtering, coding, estimating, analyzing, detecting and recognizing, synthesizing, classifying, recording, and reproducing signals by digital and/or analog devices or techniques, while the scope of "signal" includes audio, video, speech, image, musical, multimedia, data/content, geophysical, sonar/radar, bio/medical, sensation, etc. Networking suggests transportation of such multimedia contents among nodes in communication and/or computer networks, to facilitate the ultimate Internet.

Theory, technologies, protocols and standards, applications/services, practice and implementation of wired/wireless networking are all within the scope of this series. Based on network and communication science, we further extend the scope for 21st Century life through the knowledge in robotics, machine learning, embedded systems, cognitive science, pattern recognition, quantum/biological/molecular computation and information processing, biology, ecology, social science and economics, user behaviors and interface, and applications to health and society advance.

Books published in the series include research monographs, edited volumes, handbooks and textbooks. The books provide professionals, researchers, educators, and advanced students in the field with an invaluable insight into the latest research and developments.

Topics covered in the series include, but are by no means restricted to the following:

- Communication/Computer Networking Technologies and Applications
- Queuing Theory
- Optimization
- Operation Research
- Stochastic Processes
- Information Theory
- Multimedia/Speech/Video Processing
- Computation and Information Processing
- Machine Intelligence
- Cognitive Science and Brian Science
- Embedded Systems
- Computer Architectures
- Reconfigurable Computing
- Cyber Security

For a list of other books in this series, visit www.riverpublishers.com

Artificial Intelligence, Blockchain and IoT for Smart Healthcare

Hitesh Kumar Sharma

University of Petroleum and Energy Studies, Dehradun, India

Anuj Kumar

University of Petroleum and Energy Studies, Dehradun, India

Sangeeta Pant

University of Petroleum and Energy Studies, Dehradun, India

Mangey Ram

Graphic Era Deemed to be University, Dehradun, India

River Publishers

Routledge
Taylor & Francis Group

LONDON AND NEW YORK

Published 2022 by River Publishers
River Publishers
Alsbjergvej 10, 9260 Gistrup, Denmark
www.riverpublishers.com

Distributed exclusively by Routledge
4 Park Square, Milton Park, Abingdon, Oxon OX14 4RN
605 Third Avenue, New York, NY 10017, USA

Artificial Intelligence, Blockchain and IoT for Smart Healthcare / by Hitesh Kumar Sharma, Anuj Kumar, Sangeeta Pant, Mangey Ram.

© 2022 River Publishers. All rights reserved. No part of this publication may be reproduced, stored in a retrieval systems, or transmitted in any form or by any means, mechanical, photocopying, recording or otherwise, without prior written permission of the publishers.

Routledge is an imprint of the Taylor & Francis Group, an informa business

ISBN 978-87-7022-757-5 (print)
ISBN 978-10-0077-433-7 (online)
ISBN 978-1-003-33305-0 (ebook master)

While every effort is made to provide dependable information, the publisher, authors, and editors cannot be held responsible for any errors or omissions.

Contents

Preface

This book is motivated by the fact that Telemedicine and e-Healthcare have eased as well as improved the reachability of experienced doctors and medical staff to remote patients. This book presents a cross-disciplinary perspective on the concept of Artificial Intelligence, Machine Learning, Blockchain, Internet of Things (IoT), Big Data Analytics, Cyber Security, Cloud Computing, Sensors and so on that are vital to foster the development of smart healthcare and telemedicine systems.

The objective of this book is to equip the knowledge for beginners as well as for the advanced readers related to the field of smart healthcare and telemedicine. It will provide a detailed description of how advanced technologies like Artificial Intelligence, Internet of Things (IoT) and Blockchain can change the traditional way of handling patients to a smart and automated manner from remote locations. The integration of advanced techniques in healthcare can be helpful to serve humanity in a better way.

This book can be an initiator for changing the current perspective of handling patients in the traditional way physically to handling patients in a modern way remotely.

Assoc. Prof. Dr. Hitesh Kumar Sharma
Assoc. Prof. Dr. Anuj Kumar
Assis. Prof. Dr. Sangeeta Pant
Prof. Dr. Mangey Ram

List of Figures

List of Tables

List of Abbreviations

AI	Artificial Intelligence
ALS	Amyotrophic Lateral Sclerosis
ANN	Artificial Neural Networks
API	Application Programming Interface
AR	Augmented Reality
BCI	Bain Computer Interfaces
CAGR	Compound Annual Growth Rate
COPD	Chronic Obstructive Pulmonary Disease
COVID	Corona Virus Disease
CT	Computed Tomography
DLT	Distributed Ledger Technology
DNA	Deoxyribonucleic Acid
DoS	Denial of Service
ECG	Electrocardiography
e-Healthcare	Electronic Healthcare
e-Hospital	Electronic Hospital
EHR	Electronic Health Record
EMR	Electronic Medical Records
GDP	Gross Domestic Product
IoMT	Internet of Medical Things
IoT	Internet of Things
IT	Information Technology
KSI	Keyless Signature Infrastructure
ML	Machine Learning
MRI	Magnetic Resonance Imaging
NIST	National Institute of Standard and Technology
NLP	Natural Language Processing
OCD	Obsessive-Compulsive Disorde
PDSA	PLAN-DO-STUDYACT
PGHD	Patient Generated Health Data
PHI	Protected Health Information

QoS	Quality of Service
ROI	Return on Investment
SGA	System Global Area
SHA	Secure Hash Algorithm
SOA	Service Orientated Architecture
SOS	Save Our Souls
SWOT	Strengths, Weaknesses, Opportunities, and Threats
TMIS	Telecare Medical Information system
USA	United State of America
USD	United State Dollar
UV	Ultraviolet
VR	Virtual Reality
WFH	Work from Home
WHO	World Health Organization

1

Introduction to Smart Healthcare and Telemedicine Systems

Abstract

Healthcare is one of the most important sectors that needs the integration of advanced technologies. IoT-enabled sensors, Electronic Health Record (EHR) storage, advanced data analysis algorithms, metaheuristic optimization techniques are helping a traditional healthcare system for becoming smart healthcare. In a pandemic, the recent development of e-healthcare systems and telemedicine systems have proven the significance and necessity of these advancements in the healthcare sector. Smart Healthcare and telemedicine are not the solutions for all health-related problems, but these can be used for addressing many health issues without any physical movement of patients and doctors. The telemedicine system will help in the reduction of an unnecessary crowd in Private and Government Hospitals. Using these kinds of telemedicine platforms, a patient who resides in a Rural Area can get their health consultancy from various available prestigious hospitals and world-class doctors across the globe. This approach increases the efficiency and availability of an expert doctors and medical staff whenever required.

Keywords: Smart Healthcare, e-Hospital, telemedicine, EHR.

1.1 Introduction

Smart Healthcare is enablement of traditional healthcare with advanced IT technologies. It is an integration of various technologies together for providing real-time health related data gathering from a patient using smart wearable devices and diagnosis of the health issues in real-time from

1

collected data. Digital storage of health records helps patients and doctors to share information with each other from remote locations without any physical presence. It eliminates travelling time of doctors and patients, which is generally wasted in traditional healthcare approach. A telemedicine system is a smart healthcare system that is specially used for tele-based consultancy [1]. In this system, a patient sitting at their home can directly connect to the doctor remotely through a telemedicine system. The doctor can interact with the patient through audio/video call and using stored electronic health records; a doctor can diagnose the disease for the patient and prescribe him the required medicine. Telemedical information systems are playing important role in providing health services to patients. There are certain advantages of these services, it can save patient's time and expenses. The geographical distance between patient and doctor is eliminated [2, 3]. Sitting at home, a user can easily share their health information with doctors and a doctor can easily check the data, irrespective of geographical locations. Since TMIS is provided over the internet, it is vulnerable to various confidentiality and integrity attacks. It also provides privacy to patients and guarantee of the reliability of the system [4, 5]. In the Internet of Things (IoT) environment, various system and devices like embedded systems, mobile devices, actuators and sensors can receive huge amounts of information through data exchanging and inter-connection which solves the most important issue and preserve individual privacy and secure the shared data. Use of medical sensors for better treatment of patients by reaching inaccessible parts of body, minimizing the disruption of the body functions and minimizing energy consumption. These sensors can provide various functions like diagnostics, implants, electrophysiology treatments, navigation, orthopaedic aids.

Telemedicine and e-hospital already proved their existence worldwide in pandemic situations like COVID-19. This technology and various available platforms were required for remote guidance when physical presence for doctors and patients is not possible. Some of the most popular Web/Mobile based telemedicine systems are given below:

- Teladoc (US)
- Doctoroo (Australia)
- Livi (UK)
- Practo (India)
- WhiteCoat (Singapore)

In 2018, the market size of telemedicine was USD 34.28 billion. It has been projected in a survey that it will be USD 185.66 billion by 2025, having CAGR of 23.5% in the predicted timeline. As per Telehealth Index (2019) by American Well's, 350,000–595,000 US physicians will be active on Telehealth technology by 2022. As per the survey of American Well's, the following points were identified:

- 77% efficient use of the time of doctors and patients
- Reduced health care cost by 71%
- 71% effective communication between doctor and patients
- 60% enhancement in doctor and patient relation.

Telemedicine is a platform, which has Technical and Socioeconomic relevance in the Healthcare sector. Telemedicine systems have already proved their relevance across the Globe but India is behindhand to use this extraordinary and revolutionary platform because of Digital Illiteracy in Rural areas [6, 7]. Peoples living in rural areas are not aware of these lifesaving technical platforms. All developed countries in the world are using these platforms to serve high-quality healthcare facilities in minimum timespan and generate revenues to get economic benefits. It is possible for these developed countries just because of digital awareness and the proper IT infrastructure available. As I have mentioned earlier, in 2018, the market size of telemedicine was USD 34.28 billion. It has been projected in a survey that it will be USD 185.66 billion by 2025, having CAGR of 23.5% in the predicted timeline. As per Telehealth Index (2019) by American Well's, 350,000–595,000 US physicians will be active on Telehealth technology by 2022.

1.2 Traditional Healthcare vs Smart Healthcare

A traditional healthcare system is a system in which a patient moves from his residence to a hospital/clinic for his health-related issues. In some cases, the doctor needs to move to the patient's residence for diagnosis and prescribed medicines. In this approach of health treatment time and cost is unnecessarily wasted. The travelling time for patients and doctors can be easily saved with some advancement in this traditional approach.

There are various advanced technologies that may be useful to convert traditional healthcare to smart healthcare. In smart healthcare, there is no requirement for the physical movement of patients and doctors [8]. Due to this the doctors can diagnose more patients as their travelling time is completely saved and he/she can diagnose more patients at the same time.

In Figure 1.1, we have shown an example of a basic module of a smart healthcare system. In this diagram we have shown that using IoT sensors,

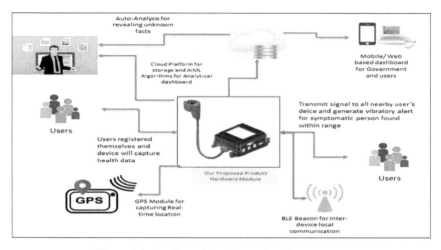

Figure 1.1   Basic module of smart healthcare system

GPS, cloud storage, AI-based data analysis, Mobile/Web based dashboard are helpful for collecting, storing and analysing the real-time data of a patient. This diagram shows a small use case of smart healthcare. In this diagram, we have shown a smart wearable device used for alerting on violating social distance constraints.

Recent technological advancements are offering users new and easier ways to access healthcare services. With the advent of high-speed networks, low-cost storage, inexpensive telecommunication systems, patient monitoring systems, cloud computing, the Tele-care Medical Information system (TMIS) is becoming a reality. Due to the advantage of telehealth care medical system, we are reaching directly to the patient's home over the internet or mobile networks and data is stored as Electronic Medical Records (EMR) [8, 9]. The major challenge here is ensuring secure access of communication data by patients and doctors, for this we need a secure and efficient way of user access control, so that attackers cannot impersonate the user or medical server. Confidentiality, integrity and availability need to be ensured. Also, the authentication process should be user-friendly so that even elderly patients can use it easily. Medical data includes medical images, which might be tampered with by intruders. One of the challenges in TMIS is to overcome the limitation of the present market scenario and provide an efficient way of communication between user and doctor with help of the cloud-based system.

Table 1.1 Comparison of traditional healthcare and smart healthcare

Traditional Healthcare	Smart Healthcare
In this healthcare system, physical movement of patient and doctor is required.	In this healthcare system, physical movement of patient and doctor is not required.
Health records and doctor prescription is stored in printed or hand written format in hospital or with patient.	Health records and doctor prescription is stored in digital format on cloud storage.
More cost and time need to be invested by patient and doctor due to unnecessary movement.	Less cost and time need to be invested by patient and doctor due to elimination of unnecessary movement because of remote consultancy.
Real-time collection of health-related data on daily basis and storage is not possible in this system.	IoT enabled wearable device helps to collect real-time data on daily basis and cloud storage helps to store unlimited data in this system.
Precision and accuracy in diagnosis and prescription is not reliable as decisions are taken on historical and non-regular data.	Precision and accuracy in diagnosis and prescription is reliable as decisions are taken on current and historical based on regular collected data.

It will use all related advanced technologies to achieve the assigned task. We have shown the basic differences between traditional and smart healthcare in Table 1.1.

1.3 Challenges in Smart Healthcare and Telemedicine

Before COVID-19 Pandemic, Online education in Schools and Work from Home (WFH) was considered a Myth and society believes that these activities are not possible without face-to-face interaction. In the same way, e-healthcare and telemedicine were also considered an impossible process. Most health service providers and patients considered that telemedicine is also a myth in the healthcare sector. However, Digital India Mission started a few years back, this association helped in a corona-virus situation and proved that telemedicine is not a myth and it played a phenomenal role in bridging some of the pain of the lockdown. Doctors are consulting patients on Mobile apps/Phone Calls. Government of India has also started web portals for telemedicine services (e.g., https://esanjeevaniopd.in/, https://ehospital.gov.in/). The citizens of India are using these facilities and getting the consultancy at their homes by good doctors across India. Due to lack of awareness and lack of infrastructure in India, Telemedicine systems still require extraordinary efforts from the Government, Doctors and technical persons for making maximum people aware of this system and spread its awareness in their nearby places, especially in Rural Areas. Security of health records saved in the form of e-record is again a major challenge in telemedicine systems [11]. Peoples are not actively using these platforms in many countries, because they are worried about stealing of their health records. Patient data is stored digitally as EHR for maintaining large data and easy accessibility from anywhere. Medical data is highly confidential and must not be tampered with because it affects the treatment given to the patient and if the data tampers, it may lead to the wrong medication. There are some technical challenges and vulnerabilities to the storage and access of EHRs in cloud databases. Major challenges include ensuring confidentiality, privacy and integrity. A more secured record storage framework with proof of concept will increase people's confidence in using these platforms [12]. Researchers proposed some approaches for ensuring secure access, exchange of medical records and securing data and transactions using blockchain technology.

The major challenges faced by doctors, patients and society in these systems are given below.

- Lack of Infrastructure for providing high-speed internet in rural/semi-urban areas.
- Lack of digital awareness about these advanced digital systems (e-healthcare and telemedicine).
- Unavailability of effective storage security mechanism for storing electronic health records.
- User authentication protocols are not secured enough as required.
- Lack of Trust building in these digital systems by doctors and patients using remote access. They still believe in physical interaction in place of remote access for a better explanation for their health issue.

These are some major challenges faced by these smart healthcare approaches. These challenges are decreasing as we are getting more reachability of high-speed internet and more users are able to use smart digital devices like smartphones. In the same way, researchers are also developing more secured protocols and authentications schemes for making these systems more secure and authentic.

The very first technology, which comes in the role while making this service for helping doctors and patients is cloud computing. Cloud computing is the platform that is the backbone of this telecare health service where our doctors and patients get their medical reports, get their updated reports whenever doctors will receive their previous reports and all the required treatment, they will get by the doctors here. This cloud is a backhand service of this telecare health services and from there only patients and doctors can easily communicate and share their reports. Cloud also ensures the confidentiality, availability and integrity of data to patients and doctors, this helps them to communicate easily and also make the data as it is, not someone who will misinterpret the patient's data. With the help of cloud-based services, if there will be a case when demand for this telecare server will suddenly arise then in this case also, the cloud can scale their services where this server is running and also whenever demand goes down, it can scale down the services. It will help us to maintain the cost of our telecare health server. Cloud computing, with its on-demand availability, helped us to reach out to customers easily.

The second vastly known technology, which comes in the role of making the telecare health services the most efficient is Machine Learning. Here, we have one issue like some of the diseases are common to some of the patients so, that's the waste of time of us and doctors also to come to reach out to the patient and treat them. Therefore, we have some machine learning algorithms by which we can make our machine intelligent so that our machine will detect that the newer patients have similar symptoms of some disease and treat them with the previous treatment given to the patients having a similar disease. It helps us to save our and doctor's time because now there is no need to treat patients with similar problems. It also makes the whole system more efficient. Machine Learning/Metaheuristic algorithms make the patient diagnosis and treatment much better because of its ability to maintain huge datasets and afterward according to the requirement came, filtering that dataset for the patient's use [13]. This ultimately led to lower cost and made the whole system powerful [14–16]. Patient satisfaction is also the prime concern for telecare health services but because of machine learning algorithms, we are ready to provide that to patients also [17]. So, this is how telecare health services play an important role in providing better services to us without even having fear of going out of our homes. In this post-COVID-19 situation, we don't now need to move here and there for our treatments as we can see there is so much hustle in hospitals for treatment, so we will be safeguarded by this telecare server as we don't need to go outside and can get proper treatment on the cloud [5]. The objective of the proposed approach is to help patients get access to health care services from home. In this approach, the patient's disease is diagnosed remotely, prescriptions and suggestions are given to the patients based on the intensity of symptoms, age and location. Also, patients showing symptoms of COVID-19 are filtered. In this model as shown in

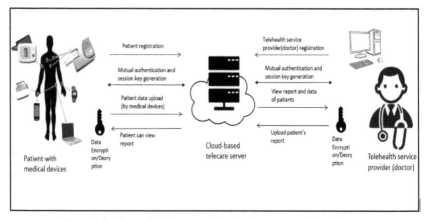

Figure 1.2   Smart healthcare authentication process flow diagram.

Figure 1.2, we have a central cloud-based telecare server, telehealth service providers and patients (or users). Patients can easily access healthcare services using any mobile device with the internet (phone, laptop, etc.,). A telehealth service provider, at any location, can be a doctor, a team of doctors, a clinic or a hospital. The approach is working in different phases as described below.

1.4 Steps Involved for Accessing Smart Healthcare and Telemedicine

There are some common steps needs to be followed by telemedicine systems are given below.

i. Registration
   a) User enters user ID, password, hospital id.
   b) This submitted data will be stored as the inactive users, passwords in hashed form.
   c) Hospital admin can activate user account of doctors.
   d) Patient's account can be activated by the doctor while uploading EHR for the first time.
   e) On account activation, user data (public key, name) are mapped in smart contract deployed on the private chain.

ii. Login
   a) Enter ID, password.

b) Verify password hash. If matched, login success else, login failure.

c) Generate symmetric key, encrypt with user's public key and store in blockchain smart contract.

iii. Give Access Permission

a) Patients can give EHR access permission to doctor by entering doctor's user id and verifying patient passwords.

b) If verification is done, decrypt symmetric key with the private key, encrypt it with doctor's public key, store in the chain, mapped with doctor's address.

iv. Update EHR

a) Check access permission.

b) If have permission, get a symmetric key from the chain, decrypt it using private key.

c) If EHR data is present then decrypt EHR with the symmetric key.

d) Match hash of EHR with the hash stored in a smart contract.

e) If hash matched, Update EHR, encrypt using the symmetric key.

f) Generate new hash and store again in a chain.

v. View EHR

a) Check access permission.

b) If have permission, get an encrypted symmetric key from a chain, decrypt it using the private key.

c) Get encrypted EHR from cloud storage, decrypt it using a symmetric key.

d) Match hash of this EHR with the hash stored in the chain. If hash matches display EHR.

After successfully registration and login by patient and doctor, both can exchange the information of digital health records i.e., EHR using the same telemedicine platform. The doctor can check patient lab reports and other health related information from these EHRs and can write a prescription to patients for their disease.

1.5 Conclusion

Smart Healthcare or telemedicine System is not the solution for all health-related problems, but it can be used for addressing many health issues without any physical movement of patients and doctors. It will help in the reduction of an unnecessary crowd in Private and Government Hospitals. Using these kinds of telemedicine platforms, a patient who resides in a Rural Area can get their health consultancy from various available prestigious hospitals and world-class doctors across the globe. This approach increases the efficiency and availability of an expert doctors and medical staff whenever required. The necessity of these kinds of systems is also required quality and safety in the same manner. These systems are directly deal with the health of the general public, so a strong and efficient authentication mechanism also need in these systems. Block-chain based authentication is a proposal that we have proposed in this chapter. Block-chain based authentication scheme can be implemented in telemedicine and it can secure this system as it secured cryptocurrency.

References

[1] R.K. Kustwar, and S. Ray, (2020). E-Health and telemedicine in India: An overview on the health care need of the people. *Journal of Multidisciplinary Research in Healthcare*, 6(2), 25–36. https://doi.org/10.15415/jmrh.2020.62004.

[2] A. Saha, et al., (2019). Review on 'Blockchain technology based medical healthcare system with privacy issues'. *Security and Privacy*. 2. 10.1002/spy2.83.

[3] S. Taneja, E. Ahmed, and J.C. Patni, (2019). 'I-Doctor: An IoT based self patient's health monitoring system', *International Conference on Innovative Sustainable Computational Technologies*, CISCT.

[4] A. Sharma, et al., (2018). Health monitoring and management using IoT devices in a Cloud Based Framework. In: *International Conference on Advances in Computing and Communication Engineering* (ICACCE), pp. 219–224.

[5] N. Kamdar, et al., (2020). Telemedicine: A digital interface for perioperative anesthetic care, Anesthesia and analgesia: 130(2), 272–275, doi: 10.1213/ANE.0000000000004513.

[6] S. Salman, et al., (2020). A secure blockchain-based e-health records storage and sharing scheme, *Journal of Information Security and Applications*, 55, 102590.

[7] H.M. Hussien, et al., (2019). A systematic review for enabling of dvelop a blockchain technology in healthcare application: Taxonomy, substantially analysis, motivations, challenges, Recommendations and future direction. *Journal of Medical Systems*, 43(10). doi: 10.1007/s10916-019-1445-8.

[8] S. Hitesh Kumar, and J. C. Patni, (2020). 'Pandemic diagnosis and analysis using clinical decision support systems', *Journal of Critical Reviews.*

[9] S. Gupta, and H.K Sharma, (2021). 'User anonymity based secure authentication protocol for telemedical server systems', *International Journal of Information and Computer Security.*

[10] Shailender, and H.K. Sharma, (2018). Digital cancer diagnosis with counts of adenoma and luminal cells in plemorphic adenoma immunastained healthcare system, IJRAR, 5(12).

[11] J.C. Patni, P. Ahlawat, and S.S. Biswas, (2020). 'Sensors based smart healthcare framework using internet of things (IoT)', *International Journal of Scientific and Technology Research* 9(2), pp. 1228–1234.

[12] S. Purri, et al., (2017). 'Specialization of IoT applications in health care industries', *Proceedings IEEE International Conference Big Data Analysis Computer Intelligence* (ICBDAC), pp. 252–256.

[13] A. Joshi, et al., (2020). 'Data mining in healthcare and predicting obesity,' in *Proceedings of the Third International Conference on Computational Intelligence and Informatics*, pp. 877–888, Hyderabad, India.

[14] G. Negi, et al., (2021). 'Optimization of complex system reliability using hybrid grey wolf optimizer'. *Decision Making: Applications in Management and Engineering*, 4(2), 241–256, https://doi.org/10.31181/dmame210402241n.

[15] G. Negi, et al., (2021). GWO: A review and applications. *International Journal of System Assurance Engineering Management* 12, 1–8, https://doi.org/10.1007/s13198-020-00995-8.

[16] A. Kumar, et al., (2018). 'Complex system reliability analysis and optimization,' *Advanced Mathematical Techniques in Science and Engineering*, 1, 185–198, River Publishers.

[17] D.S.R. Krishnan, et al., (2018). An IoT based patient health monitoring system, 2018 *International Conference on Advances in Computing and Communication Engineering* (ICACCE), p. 1–7.

2

Advanced Technologies Involved in Smart Healthcare and Telemedicine Systems

Abstract

According to a comprehensive study by market and market, the global health-care market is estimated to reach more than USD 829 billion by the year 2026 from USD 319 billion in 2021 at a compound annual growth rate of 21% during the period under consideration. Such a humongous growth of the healthcare market is attributed to the changing policies of governments where many of them are investing highly in healthcare, telemedical and IT solutions to existing problems. This growth can also be attributed to the high ROI and business growth associated with healthcare IT solutions, the accumulation of big data and the escalation in its use for devising better solutions, the demand to curb the rising prices of healthcare and many more.

In this chapter, we have described the involvement of some advanced technologies like robotics, artificial intelligence (AI), the Internet of Things (IoT), blockchain, cloud computing, etc. in smart healthcare and telemedicine.

Keywords: smart healthcare, e-hospital, telemedicine, cloud computing, artificial intelligence, IoT, blockchain.

2.1 Introduction

With the aid of modern technologies, smart healthcare is able to provide high-quality solutions in the field of healthcare and medicine. Smart healthcare becomes a new reality and this original idea provides incredible service to COVID-19 patients and executes exact operations. The present pandemic situation is easily managed and digitally controlled due to the modern technologies associated with smart healthcare. In the world of medicine, advanced technologies take on new problems in creating effective support systems for physicians, surgeons and patients in the medical world. Information technologies and electronics play a vital role in controlling and managing the current pandemic situation. In India e-health and telemedicine is in the first phase [1]. It needs more development and investment for smart healthcare and telemedicine.

Some of the advanced technologies are listed below and the involvement of these latest technologies can help to convert traditional healthcare into smart healthcare.

- Artificial intelligence and machine learning
- Robotics
- Cloud computing and storage
- Cyber security and forensic
- IoT
- Blockchain
- Computer vision

In his research work, Saha et al., (2019) have defined the use of blockchain in telemedicine [2]. Various authors have proposed IoT based systems or cloud-based systems for smart healthcare systems [3–5].

In this chapter, we have given a detailed description of some of these technologies applications in smart healthcare.

2.2 Advanced Technologies Used in Smart Healthcare

2.2.1 AI in Smart Healthcare

AI works with the help of data from the past to make a decision as accurate as possible for the future [15]. A common example of AI is our E-mail. We have different labels for different types of mails. The AI algorithm asks users to label some mails as primary, social, promotions and updates. Now on basis of the user input, it automatically categorizes future mails into different labels on its own.

Some applications of AI in healthcare are given below (Figure 2.1):

- **Wellness monitoring:** In the 21st century we have developed wearable gadgets like smartwatches which monitors our heart rate, SpO2 levels and even our stress levels using AI to ensure our health and wellness [6, 7].

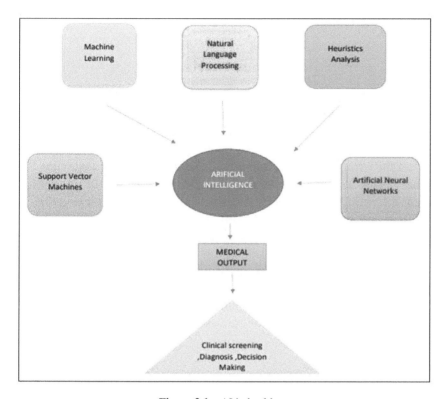

Figure 2.1 AI in healthcare

- **Increased efficiency:** In some cases, it was seen that the machines using AI showed more accurate and fast results as compared to other methods used to detect X-Rays, CT-Scans, etc. [8]
- **Surgical robots:** One of the most useful and life-changing uses of AI in healthcare is through surgical robots. Nowadays, these robots assist doctors [9]. But, in future they might replace the need for a doctor during a mild operation.
- **New pharmaceutical drug discovery:** Based on the existing knowledge of drugs that we have today the machines scan new combinations and patterns to find out a new pharmaceutical drug.
- **Decisions based on data:** The AI works in such a way that it eases doctor's decision-making. If the doctors get stuck at some point during an emergency, they can use AI to somewhat detect the consequences of their next step.

Some challenges faced by AI in healthcare are given below:

- **Data collection:** As we know that AI takes data goes through that data and take decision based on data. So, it is very important to provide the right data otherwise the decision would cause catastrophe.
- **Training staff:** Not everyone in the staff may be doctors, nurses or helpers might be a techie. They might fail to handle the systems, which could manipulate the results.
- **Considering failures:** The artificial intelligence technology must also consider the wrong decisions from the data. It should not use something that has caused some problem to the patient in the past.
- **Patient's trust:** One major thing that artificial intelligence is lacking is trust. If we ask patients and their family members to get operated on a robot most of them would not trust it and would refuse to have surgery.

2.2.2 Cloud Computing in Smart Healthcare

Cloud Computing is a domain in IT Industry that provide all IT resources online. The user need not buy physical hardware resources for computation and storage. In smart healthcare and telemedicine system, there is a requirement of large storage for EHR (Electronic Healthcare Records) and a fast-processing platform for analysis.

Hence, there is a need of using Cloud computing services in Smart Healthcare for fast processing, large storage and high availability.

2.2.3 Cyber Security and Forensic in Smart Healthcare

This domain of Information Technologies deals with the security and safety aspect of digital records stored on Cloud storage of smart healthcare and telemedicine systems. Being a fully network-controlled system Security and tolerance towards network attacks is one of the major requirements for smart healthcare systems.

Designing effective security protocols, providing a highly secured authentication scheme is a core responsibility of this domain of Information technology.

2.2.4 IoT in Smart Healthcare

A patient's interaction with a medical practitioner is largely limited to the appointments with them [10]. Without the use of sensors or special monitoring devices, there is no way of continuously monitoring the patient's health stats and recommending the necessary actions [11].

The IoT has made a revolution in many different fields and healthcare is no different (Figure 2.2). This has brought home the idea of remote monitoring and remote sensing of the health status and other vitals of a patient, allowing medical experts to take actions immediately and on the fly. Usage of sensors and other IoT devices in conjunction with cloud and big data services has enabled technologists to devise some incredible solutions that not only reduce the overall costs of healthcare but increase the expectancy of life, provision better treatment opportunities and significantly increase the overall satisfaction of the patients. With the ever-increasing fast-paced life, continuous monitoring and alerting systems are required. Modern machinery capable of reacting to changing stimuli is becoming the crying need of the hour. IoT brings forth solutions that are capable of changing and adjusting to changing stimuli and generating appropriate signals when needed.

Today, smart TV, smart Watch, AI-based camera sensing, AI-biometric, smart refrigeration for sensitive chemicals, drugs and specimens, etc all

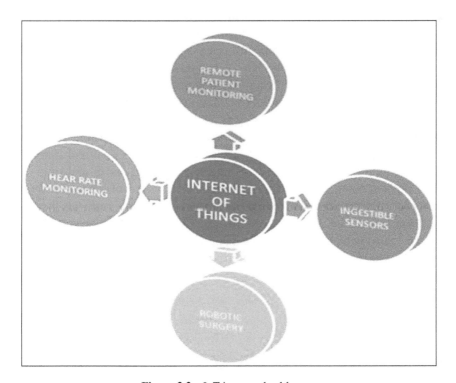

Figure 2.2 IoT in smart healthcare

require dynamic systems [12, 16]. IoT helps them in it. In healthcare, measurement of health vitals on a regular basis and then changing the required plan of action according to the vitals recorded, continuously monitoring the internal functioning and sophisticated measurements taken by precise instruments in hospitals and clinics, safe and robust delivery of sensitive drugs in environment-controlled environments, etc. all these problems can be solved via IoT. These issues if not solved can cause losses of billions of dollars every year.

- **IoT device for monitoring vitals:** These days' fit bits have become extensively popular. Similar and better IoT-based devices that are wearable and that can monitor vital health stats of a patient including blood pressure, heart rate, glucose levels, thyroid levels, etc can be used for personalised and customized care for patients. Devices like these keep a check on the patient and can be used to alert the required doctor or nurse in case needed. These devices will also be able to furnish continuous data of the health vitals which will ultimately lead to better patient analysis.

- **IoT in hospital equipment:** Medical equipment attached with sensors and other types of monitoring and tagging devices will help better control and regulate the equipment even from a remote location. This ensures that the medical equipment keeps running hassle-free and yet be monitored and supervised even when nobody is physically present. Heavy duty and expensive pieces of equipment including defibrillators, oxygen pumps, heart rate monitoring systems, IV pumps, dialysis equipment and so on and so forth can be given an additive attachment of these IoT devices which will enable professionals to track their real-time activity and location and take immediate necessary actions in real-time Environment controlled chambers. Storage facilities can also be created using IoT devices. This will help preserve specimens and chemicals that would otherwise be destroyed by the unsanitary or harsh environment, thereby causing humongous loss of manpower and money.

- **IoT for pharmaceutical companies:** Companies dealing with drugs and their delivery and storage can use IoT devices for controlled environment storage and delivery of drugs/specialized chemicals. Certain pharmaceutical drugs or chemicals are extremely volatile or sensitive hence they can be damaged by even the slightest changes in temperature or humidity. IoT devices keeping a regular check on such environments create a safe zone for their storage and transportation.In addition, IoT-based tracking of equipment is necessary. This is particularly helpful when avoiding counterfeiting and more agile delivery.

Advantages of IoT in health industry

- **Remote tracking:** Real-time remote tracking using connected IoT devices and smart indicators can detect illnesses, treat illnesses and save lives via sending SOS signals in the event of a medical emergency.
- **Reduced healthcare costs:** For the matter of some mild medical condition such as fever, infection, allergies, etc., one do not have to visit a doctor physically. Rather one can sit at home and have a hassle-less consult with a doctor.
- **Cost-cutting/cost saving:** After the implementation of IoT systems in hospitals both, the customer (patient) and the hospital/Doctors can save their cost of operations. The patient now does not have to travel miles to reach a good doctor. The doctor can now also function from his/her home reducing their cost of operation, which in turn would maximize the profit.

Disadvantages of IoT in health industry

- **Privacy:** With the advancement in technology these days, there are an increasing risk of data getting stolen, systems getting hacked and much more, which can be used against the consumers.
- **Expensive:** The initial amount required to set up IoT systems, machines are really high which could be a turn-off for a lot of hospitals/organizations.
- **Mechanical/software failure:** A small failure in the machinery or the software system can cause someone's life. The data needs to be monitored accurately.

Even though IoT-based solutions bring home a plethora of advantages, the cost of building them is generally high. So many times, it might not be possible. Overall, with the reduction in the prices of hardware every day and in the light of innovation, IoT-based solutions will make the healthcare industry much better.

2.2.5 Blockchain in Smart Healthcare

Healthcare, like any other industry, are extremely data intensive. Huge amounts of data is generated and consumed every day. The supply chain of pharmaceutical devices, drug delivery systems, health record management, patient data management, etc., all have been run via un-automated computer

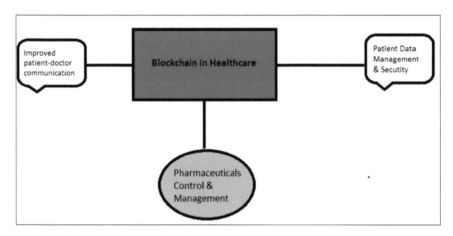

Figure 2.3   Blockchain in healthcare

systems and need tedious paperwork. This kind of existing system is the root cause of manual and administrative errors, flaws in the delivery pipeline, loss of man-hours and money and increase in the overall costs. With these traditional systems, becoming increasingly unreliable and costly, new solutions to the problems at hand are the need of the hour. Here, Blockchain can be used to provide an up-and-coming solution to existing issues [13, 14]. The healthcare industry suffers from mundane and rigorous manual tasks that are cost-inefficient and time-consuming. Approximately thirty billion transitions happen every year in the healthcare industry, of which about fifteen billion are manually faxed, costing the industries involved in it, USD 250 billion. That's not all, even with the cost being exceptionally high, more than 60 % of the medical practitioners do not receive adequate and correct information or in time. Miscommunications, delays and administrative errors in the medical and healthcare processes and drug or patent-related data cause about 400,000 deaths every year. On top of that, with professionals filling more than 20,000 forms every year, each costing about USD 20, costs in the industry surges even more. Manual documentation is not only costly but highly prone to errors. In the following figure (Figure 2.3), we have shown some improvement areas in healthcare using Blockchain.

Transparency is almost negligible when considering supply chains or record management. Humongous amounts of data shared across parties and its inadequate, inefficient and unreliable management along with the opaqueness in the pipeline of supply chains which have become a breeding ground for drug counterfeiting, results in a loss of more than 200 billion dollars every year in the U.S.A. All these problems combined make up for a significant loss of life and money. Blockchain, as a technology, can solve a lot of these issues by making existing processes more efficient and minimizing losses. Being transparent, immutable and distributed, blockchain solutions can help reform and modernize the healthcare system.

- **Supply chain pipelines:** In the pharmaceutical/drug or equipment delivery chains, the entire process can be made more streamlined and highly efficient if blockchain is used. Blockchain can reduce the problem of counterfeit products and make transactions more robust and transparent. It will not only recuse frauds and cut losses but also increase efficiency and overall profits.
- **Procuring traceability and authenticity of products:** Blockchain is built on the foundations of transparency and security. Any organisation willing to be a part of the supply chain and sell it's products must be trustworthy enough to be allowed by the authorities controlling the blockchain-based system, to list their products. Once they do so, all the data will be readily available and cannot be changed. Furthermore, time stamping and real-time tracking enables a check on fraudery and counterfeiting. Transactions becomes seamless and the authenticity of products being exchanged become more apparent.
- **Data and record management:** Data management, in itself, is a tedious task. Mixed with high importance data involving medical records, inefficient dealings with the data, multiple records of the same data being rendered by different parties using the same process, etc. will render the management portion of the data useless. Lack of data sharing, absence of authenticity and originality of data and unavailability of secure and immutable storage causes huge problems in the healthcare industry. Blockchain based record management and data organisation can relieve us of many of these problems. With data security, immutability and interoperability built into it, blockchain ensures no data breaching and data infection, thereby ensuring robustness and correctness.

Even though blockchains are a great solution to many healthcare-based problems, these solutions are not perfect [9]. It comes with its own set of challenges. Blockchain requires expensive hardware and machinery to be put in place. It requires fast machines for efficient computation so that data availability is high and latency is low. Apart from that, the data that each block in a blockchain can store is quite low. In addition, because medical data is generally quite large, it poses a challenge to manage ever-increasing quantities of data. Overall, efforts can be made to ensure working solutions for existing problems. Blockchain is a commendable candidate for building top-notch modern solutions to problems that affect an industry dealing in billions of dollars. Blockchain is an online ledger system that stores data/information in such a way that it can't be edited, changed or hacked. It is a very safe and latest technology, which has a good future ahead. Since this technology needs to tackle important data, it has been created in such a way that it cannot tamper, with changed or deleted using some outside sources.

Following are some challenges for IoT in smart healthcare:

- **Data blockage:** Let's say a person goes to hospital 'A' every time for their health checkups. Now, hospital 'A' has all the reports and records of the person. But, in a case of emergency if he/she is rushed to hospital 'B'. When 'B' asks for the data and 'A' refuses to share the data.
- **Data corruption:** Consider a situation where the data stored with the hospital or healthcare centre gets corrupted for some reason. Now the data is no more with the hospital.

However, there are ways to change the data but it is secure. Since it is a new technology, we can trust the technology.

2.3 Conclusion

Information Technology gave revolution in every sector of industry and society. Healthcare is also a core sector, which needs advancement. Integration of the latest IT-based technologies in healthcare will help it from becoming a smart healthcare system. AI, IoT and blockchain are some of the advances IT based technologies that have already started their involvement in the healthcare sector. Telemedicine and e-hospital are a reality now and patients are connecting to the doctor using these online platforms and getting prescriptions for these diseases without any physical movement.

References

[1] K. Kustwar, and R. Suman. (2020). 'E-Health and telemedicine in India: An overview on the health care need of the people.' *Journal of Multidisciplinary Research in Healthcare*, 6(2), 25–36. https://doi.org/10.15415/jmrh.2020.62004.

[2] A. Saha, et al., (2019). Review on 'Blockchain technology based medical healthcare system with privacy issues'. *Security and Privacy*. 2. 10.1002/spy2.83.

[3] S. Taneja, et al., (2019). 'I-Doctor: An IoT based self patient's health monitoring system', *International Conference on Innovative Sustainable Computational Technologies*, CISCT.

[4] A. Sharma, et al., (2018). 'Health monitoring and management using IoT devices in a Cloud Based Framework.' In: 2018 *International Conference on Advances in Computing and Communication Engineering* (ICACCE), pp. 219–224.

[5] N. Kamdar, et al., (2020). Telemedicine: A digital interface for perioperative anesthetic care, Anesthesia and analgesia: 30(2), p. 272–275, doi: 10.1213/ANE.0000000000004513.

[6] K.S. Hitesh, and J.C. Patni, (2020). 'Pandemic diagnosis and analysis using clinical decision support systems', *Journal of Critical Reviews*.

[7] S. Gupta, and H.K. Sharma, (2021). 'User anonymity based secure authentication protocol for telemedical server systems', *International Journal of Information and Computer Security*.

[8] Shailender, and H.K. Sharma, (2018). 'Digital cancer diagnosis with counts of adenoma and luminal cells in plemorphic adenoma immunastained healthcare system,' IJRAR, 5(12).

[9] A. Joshi, et al., (2020). 'Data mining in healthcare and predicting obesity,' in *Proceedings of the Third International Conference on Computational Intelligence and Informatics*, pp. 877–888, Hyderabad, India.

[10] J.C. Patni, et al., (2020). 'Sensors based smart healthcare framework using internet of things (IoT)', *International Journal of Scientific and Technology Research* 9(2), pp. 1228–1234.

[11] S. Purri, et al., (2017). 'Specialization of IoT applications in health care industries', *Proceedings IEEE International Conference Big Data Analysis Computer Intelligence* (ICBDAC), pp. 252–256.

[12] D.S.R. Krishnan, et al., (2018). 'An IoT based patient health monitoring system,' *International Conference on Advances in Computing and Communication Engineering* (ICACCE), p. 1–7.

[13] S. Salman, et al., (2020). A secure blockchain-based e-health records storage and sharing scheme, *Journal of Information Security and Applications*, 55, 102590.

[14] H.M. Hussien, et al., (2019). A systematic review for enabling of develop a blockchain technology in healthcare application:

Taxonomy, substantially analysis, motivations, challenges, Recommendations and future direction. *Journal of Medical Systems*, 43(10). doi:10.1007/s10916-019-1445-8.

[15] G. Negi, et al., (2021). Optimization of complex system reliability using hybrid grey wolf optimizer. Decision making: *Applications in Management and Engineering*, 4(2), 241–256.

[16] P. Koranga, et al., (2018). Image denoising based on wavelet transform using visu thresholding technique. *International Journal of Mathematical, Engineering and Management Sciences*, 3(4), 444–449.

3

Role of Artificial Intelligence, IoT and Blockchain in Smart Healthcare

Abstract

Information Technology has shown his presence in every sector which require automation and intelligence. Tradition healthcare is also such a major sector in which lots of advancement is needed. AI, IoT and Blockchain are the three mains IT-based advanced technologies that are required at many phases of converting traditional healthcare to smart healthcare.

In this chapter, we have described the involvement of some advanced technologies like artificial intelligence (AI), Internet of Things (IoT) and blockchain in smart healthcare and telemedicine.

Keywords: smart healthcare, e-hospital, telemedicine, artificial intelligence, IoT, blockchain.

3.1 Introduction

The three major requirements of a smart healthcare system are data collection; data analysis and data security and all these three requirements can be fulfilled by IoT, AI and Blockchain technologies. The use of AI-based algorithms and software to replicate human cognition in the analysis, display and comprehension of complicated medical and healthcare data is referred to as AI in healthcare. AI is defined as the ability of computer algorithms to make educated guesses based purely on input data. AI can be utilised to perform the same tasks in a more efficient and cost-effective manner. It is always preferable to prevent than cure. Artificial intelligence-based apps can assist users in leading a healthy lifestyle and being proactive. When customers realise, they have power over their own health, they are more motivated to live a healthy lifestyle.

Remote surveillance in the health sector has been made possible by IoT-enabled technologies which unleash the potential to safely and healthily maintain patients and empower doctors to provide exceptional treatment. The facilitation and efficiency of contacts with clinicians boosted patient participation and satisfaction. In addition, remote health monitoring help reduces hospital stay time and reduce re-admission. IoT offers healthcare, family, doctors, hospital and insurance businesses. In IoT wearables and other home-monitoring technology included in IoT can help doctors track patients' health more effectively. Anyone can monitor patients' adherence to treatment regimens or any immediate need for medical care. IoT assists healthcare professionals in being more attentive and proactive with patients.

The challenge of trust in a complicated setting is well solved by Blockchain. We will give an example: at first, this technology and cryptocurrency was made viable since their use did not necessitate the involvement of a centralised body. Previously, if we disagreed with the system's operation, we went to the bank and more often than not, we just trusted it because the government was behind it. The entire practice of money exchange was based on this, for the most part. We were able to construct a decentralised system with a set of rules that everyone agreed on and that is very tough to deceive using the Blockchain. The capacity to automate operations via smart contacts is the second thing the Blockchain excels at. Even this, though, is a contentious topic. Numerous technologies aid in the automation of operations none of which are better or worse than the Blockchain. Initially, this technique was used in the financial sector. Even though not everyone understands how it should be used in this area, there is currently a mass of at least half of the finished products.

3.2 Role AI in Smart Healthcare

All of us have been witnessing drastic technological improvement since the 20th century. From a time when we used to think about self-driving cars, we now have access to the autopilot-driving feature in Tesla. We get personalized feed on all our social media handles, such as YouTube or Instagram. Various voice assistants such as Siri, Alexa have been made available to us. They perform all the required tasks as soon as we instruct them. These are a few examples from our daily lives where we can understand the use of AI and data mining techniques [1]. Sometimes our lives are highly intertwined with use cases of AI and we don't even realize it. The definition acts as a self-explanatory statement that implies that once we teach a machine to solve a particular problem, it will become capable enough to solve all other problems of a similar pattern and gradually improve itself. The word intelligence in AI should not be mistaken with the machine performing something very creative or clever. Using AI, the machine can assist doctors and pathologists to diagnose a disease like cancer [2, 3]. It simply means that machines are trained to do the same tasks that are performed by humans and thus simulate human behaviour. The term intelligence consists of a few factors:

1. **Generalization learning:** This implies that users will be able to perform better even during situations that have not been encountered earlier.
2. **Reasoning:** This implies that the user will be able to draw conclusions for the provided problems.
3. **Perception:** Analyzing features and relationships between objects.

Since the 1950s, this field has undergone extensive research. This technology has certainly benefited people by solving some major issues that society was facing. It has also paved the way to a lifestyle where people can use machines to simplify routine tasks. There is no end to how AI is being used in various sectors. Banks are using this technology to create chatbots that can interact with customers 24/7 to solve their queries [4]. This method eliminates the bound time frame that is followed by the traditional banking approach. Many cases of the robotics field using AI to perform mundane tasks or tasks are deleterious for humans. There are many instances where AI is used in the healthcare industry. For instance, various surgeries now are AI-assisted. Drug discoveries that use deep learning are very precise and require considerably fewer resources like time and money. Early detection of contagious diseases has been possible due to machine learning. Thus, it is rightly said, 'Machine intelligence is the last invention that humanity will ever need to make'.

As mentioned, machine learning has been used in the domain of healthcare to detect diseases at an early stage. In addition, deep learning has been used for discovering medicinal drugs with comparatively fewer resources [15]. But what exactly are machine learning and deep learning? People generally tend to use these three terms interchangeably. So how are they related to artificial intelligence? Machine learning is a concept in which machines are not manually programmed, instead they gradually learn from the data and past experiences. It is a subset of AI and acts as the basis for data science. Machine learning models are being widely adopted in various fields such as technology, science, etc. One main reason for the introduction of machine learning was that people in some sectors were facing issues.

Researchers in the field of neuroscience faced problems while designing operation models of the brain. Secondly, it received widespread acceptance because it makes decisions based on data provided to the machine. This data acts as evidence for future reference. Deep learning in essence is a neural network that consists of various layers. It requires a large amount of data to learn and simulate the behaviour of the human brain and then make intelligent decisions and conclusions on its own. As machine learning is a subset of artificial intelligence, deep learning is further a subset of machine learning. This brings us to the conclusion of a hierarchy wherein AI can be imagined as a circle inside which we have a circle that represents machine learning. Further, we have another smaller circle inside machine learning that represents deep learning.

Technology has been the base for rapid development in all spheres. Even when it comes down to medicine and healthcare, it is believed that technology would fundamentally change the way we receive and deliver care. To understand this, consider this example of the number of diabetic patients around the globe. So according to data, there are 415 million diabetic patients in this world, out of which 62 million people reside in India. Diabetes further causes the person to be at risk for diseases, such as heart disease, kidney disease or vision loss. As far as the problem of vision loss due to diabetes is concerned, there are only 15,000 eye doctors for 62 million diabetic patients. Due to such a small number of eye specialists, about half of the diabetic people suffer from some form of vision loss before they are diagnosed, even when the disease is preventable. This situation could entirely change once AI enters the picture. Through AI, algorithms are developed that will automatically detect eye disease from the pictures that are taken. This will solve our problem of millions of patients relying on few specialized eye doctors. This technology can be used in every part of our country. This implies that people living in rural areas who do not have access to specialized healthcare will also be able to take benefit from this initiative. Therefore, people will not have to travel long distances and sacrifice their work for receiving basic health needs. Apart from these problems, AI can also help people with low hearing or people who do not speak the same language as the doctor. This can be done by translating and documenting all that has been said by the doctor in their native language.

One more example where AI is a cut above the traditional methods is for detecting cancer. AI can be used to detect lung cancer and breast cancer with an accuracy that meets or even beats the accuracy obtained by radiologists. These examples give us an idea that AI can revolutionize the industry of healthcare and benefit billions of people throughout the globe.

These examples give us a glimpse of how powerful AI can be if utilized efficiently. So now let us understand a few use cases in detail wherein AI is already in use in the field of healthcare and what changes have been observed since then. AI is making a record-breaking work in many sectors like industries, business and many more and now has potentially proved itself even in the Healthcare sector [13, 14]. Imagine when we would be able to analyze every data of patient like their visits to clinic, medications, tests, their procedure as well as data outside health like information about credit cards, etc. it would be way easier to guide patient. AI is using computers to carry out model the intelligent behavior with minimum human interactions. It is capable of solving problems, writing and learning like a human brain. AI works on natural language processing (NLP) by recognizing the pattern then it makes the visual perception and makes the decision. Also, machine learning (ML) is a developing area to automatically sense data.

Advantages in machine:

1. It helps to retain and store the information.
2. Innovating into smarter and updated over time.
3. Helps to low-rise the working burden and many more.
4. The application of AI in medicinal purposes has two main branches:

    4.1 Virtual branch.
    4.2 Physical branch.

AI: The Virtual Branch

The virtual branch is completely based on Machine-Learning also called Deep Learning. It is an algorithm that improves learning through experience.

Benefits of AI are:

1. **1.** AI can assist doctors and physicians by:

    1.1 Making better clinical decisions.
    1.2 Helps in replacing humans in certain functional areas like radiology.
    1.3 Updated medical info.
    1.4 24/7 expert availability.

2. **2.** Effective in early diagnosis of diseases.
3. **3.** Predicting the outcomes and treatment of the diseases.
4. **4.** Patient's feedback after the treatment that is very helpful as well as important.
5. **5.** Strengthen and supports non-pharmacological management.

6. Reduces errors in diagnosis and treatment.

7. Increases safety of patients and saves their savings

8. Has learning ability and self-updating with the given feedback.

AI: The Physical Branch

It includes many Physical objects, Medical devices, carebots (robots to take care of)/or robots in surgeries.

In Figure 3.1, we have shown the cyclic process of AI in a AI-enabled healthcare system.

Even though there are various challenges that are to be faced by AI such as Development costs, Integrational issues such as Ethical issues and non-satisfactory results by medical practitioners in adopting innovative and new technologies in AI and the most importantly fear of replacing humans with a device. It may face issues like Data privacy and security in devices that we use for AI. It may need some continuous training to get some brief and generalize the idea of the system but studying it in brief on both positive and negative sides convinces us that AI will prove to be a boon to our healthcare.

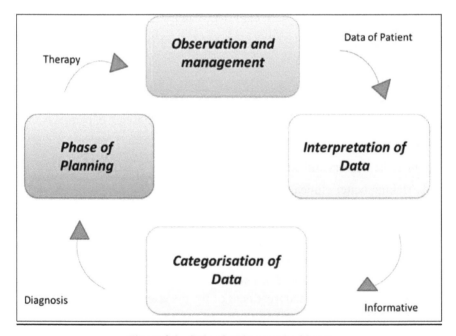

Figure 3.1   Role of AI in smart healthcare

3.3 Role IoT in Smart Healthcare

IoT in Healthcare means getting connected to devices in hospitals, clinics and many other medical settlements which include sensors like thermostats, automated lights, security [5]. It may refer to medical devices consolidated into frameworks. Medical settlements that include sensors like thermostats, automated lights, security. It may refer to medical devices consolidated into frameworks.

COVID-19 has a huge impact on life that we may never return to old normal it might be technologically or in any perspective of life. In healthcare we have been wearing various devices to make our body free or updated with all the body systems then it might be in the form of watches, belts, mobile phones, apps and many devices which we were yet to be known. In Figure 3.2, we have shown the role and cyclic process of IoT in smart healthcare.

Examples of IoT in healthcare are:

1. Remote monitoring of the patient. [6]
2. Monitoring glucose, heart rate, various diseases.

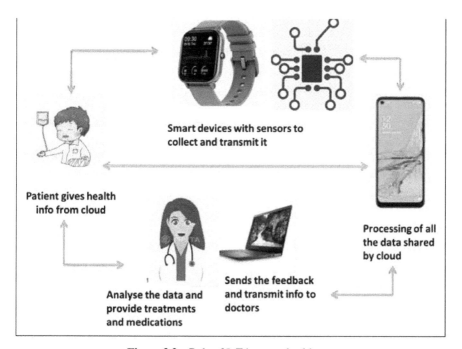

Figure 3.2   Role of IoT in smart healthcare

3. Helping in COPD by connecting Inhalers.
4. Connecting contact lenses.
5. Robotic surgeries and many more.

Talking about the impact on society IoT even has been beneficial for our society in many ways and made a positive impact on us in the following ways:

1. IoT Security has increased the safety for the patient as well as all the working staff in medical settlements.
2. IoT healthcare has made an enormous difference in precision and proper treatment for a problem by innovating wearable technology. [7]
3. Made patients way more comfortable and convenient and made them satisfactory which made them recover at a faster pace.
4. Even UV sanitation has helped a lot.
5. Made the bond of privacy and trust issues way stronger.
6. Made processes less time-consuming and facilitated doctors with decision making and made specialized workforce less burdening.

Although installing IoT to hospitals takes some of our efforts by investing initially then associating with implementing IoT infrastructure, installing devices and training staff members on the system [7], but it is going to be beneficial for the future. Mainly when IoT devices are installed it makes the difference in updating and installing new ones which make difference in non-IoT hospitals in relaunching them.

Observing all the beneficial sides of IoT it has always been said that 'The Coin has two sides' we need to over through the challenging side of IoT [8]. IoT with cloud computing can increase the data acquisition and storage efficiency of these systems [9]. Healthcare must recognize them and face when implementing it such as:

1. In security, a large healthcare network often faces problems in managing security in repositories of data and different facilities.
2. In adopting a new IoT framework is one of the biggest challenges. Moving a whole facility to a new system and then processing over a data. The initial investment, the installation process can make it difficult.
3. Data collection may vary according to the priorities of IoT system's data collection techniques and usage.

Making the full sketch of all the upcoming and ongoing studies of the IoT system in healthcare the future might be the brightest and the latest version of the present situation. It would continue to evolve and make the system much more advanced, with better outcomes and experience. We would gain many new insights about the medical environment its treatment. Many futuristic things would be innovated and invented by the will and fruitful outcomes of research. That means the IoT in healthcare is just a beginning, and many more inventions are on the horizon.

3.4 Role Blockchain in Smart Healthcare

As a strategic tool and raising a hand towards strengthening all the operative protocols and creating a proper efficient decisional process, blockchain has been applied in healthcare management. When using blockchain in combination with AI, it makes us all easier to get over this pandemic risk. A SWOT analysis in healthcare for the adoption of blockchain has contributed in it [10]. Hoping for this so that it may help us improve from the worst situation of COVID-19.

The National Institute of Standard and Technology (NIST) defines it as tampering evident and resistant digital ledgers in a distributed fashion. Under a community, it records all the transactions, which cannot be changed once published.

In Figure 3.3, we have shown the role and users IoT enabled smart healthcare system. The biggest myth is that it is tied to Cryptocurrencies [11]. Blockchain is the backbone of crypto, which turns their creation and helps in exchange through secure transactions. Cryptocurrency is digital money or u may call a medium of exchange just as a governmental currency as Dollars ($) and Rupees. Blockchain can be hacked which goes to 51% [12].

Now after a quite brief scenario and its basic understanding, it is essential to get into healthcare where all the technology there is valuable and relevant. The data stored in this chain is of a specific case. According to blockchain in healthcare should maintain the information about individual's information, their identity and information about payment cards. In many servers' data are of two types:

On-chain:

1. High-level data: It includes transaction data and records and pointers.

Off-chain:

1. Large files data: It keeps a large volume of clinical information in a secure manner.

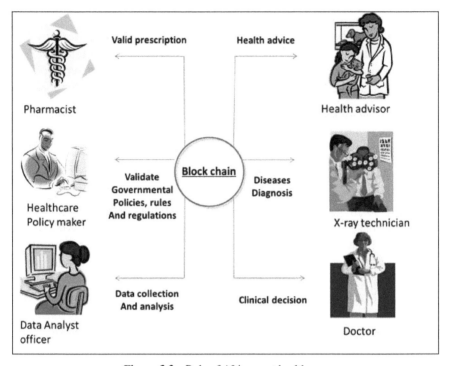

Figure 3.3   Role of AI in smart healthcare

After a detailed outline of blockchain it has proved that the whole healthcare needs systematic blockchain management for successful and frictionless continuity of healthcare. Even maintaining a detailed and smarter loop-less blockchain is the most important part.

3.5 Conclusion

These use cases have been a breakthrough in itself, but there is much more to come. Especially if people can learn about the current advancements and then think of ways in which this technology can further be applied to things that others have not even thought of as of now. This thought process and way of working will lead us to the next evolution of where we are going and for what we are trying to achieve through this. We can pursue the avenue of collecting information to prevent an adverse outcome from happening. Thus, the collaboration of doctors, engineers and scientists can pave a way for the healthcare industry where everyone will have happier and most importantly, healthier lives.

In this chapter, we have described in detail three major technical areas (AI, IoT and Blockchain) which can be useful in smart healthcare. Effective integration of these technical areas will help smart healthcare to serve humanity in a better way.

References

[1] A. Joshi, et al., (2020). 'Data mining in healthcare and predicting obesity,' in *Proceedings of the Third International Conference on Computational Intelligence and Informatics*, pp. 877–888, Hyderabad, India.

[2] Shailender, and H.K. Sharma, (2018). 'Digital cancer diagnosis with counts of adenoma and luminal cells in plemorphic adenoma immunastained healthcare system,' IJRAR, 5(12).

[3] J.C. Patni, et al., (2020). 'Pandemic diagnosis and analysis using clinical decision support systems', *Journal of Critical Reviews*.

[4] N. Kamdar, et al., (2020). Telemedicine: 'A digital interface for perioperative anesthetic care, Anesthesia and analgesia,' 130(2), p. 272–275, doi: 10.1213/ANE.0000000000004513.

[5] S. Taneja, (2019). 'I-Doctor: An IoT based self patient's health monitoring system', *International Conference on Innovative Sustainable Computational Technologies*, CISCT.

[6] J.C. Patni, (2020). 'Sensors based smart healthcare framework using internet of things (IoT)', *International Journal of Scientific and Technology Research* 9(2), pp. 1228–1234.

[7] D.S.R. Krishnan, et al., (2018). An IoT based patient health monitoring system, *International Conference on Advances in Computing and Communication Engineering* (ICACCE), p. 1–7.

[8] S. Purri, N. Kashyap, et al., (2017). 'Specialization of IoT applications in health care industries', Proc. *IEEE International Conference Big Data Analytics Computer Intelligence* (ICBDAC), pp. 252–256.

[9] A. Sharma, et al., (2018). Health monitoring and management using IoT devices in a Cloud Based Framework. In: 2018 *International Conference on Advances in Computing and Communication Engineering* (ICACCE), pp. 219–224.

[10] S. Salman, et al., (2020). A secure blockchain-based e-health records storage and sharing scheme, *Journal of Information Security and Applications*, 55, 102590.

[11] H.M. Hussien, et al., (2019). A systematic review for enabling of develop a blockchain technology in healthcare application: Taxonomy, substantially analysis, motivations, challenges, Recommendations and future direction. *Journal of Medical Systems*, 43(10). doi:10.1007/s10916-019-1445-8.

[12] S. Gupta, and H.K. Sharma, (2020). 'User Anonymity based Secure Authentication Protocol for Telemedical Server Systems', *International Journal of Information and Computer Security*.

[13] N. Uniyal, et al., (2021). Nature inspired metaheuristic algorithms for optimization, Meta-heuristics optimization techniques, Walter De Gruyter, 1–10.

[14] A. Kumar, (2018). Complex system reliability analysis and optimization. *Advanced Mathematical Techniques in Science and Engineering*, 1, 185–198, River Publishers

[15] T. Priyanka, et al., (2021). Deep learning for satellite-based data analysis. *Meta-heuristics Optimization Techniques*, Walter De Gruyter, 173–188.

4

Application of Artificial Intelligence in Smart Healthcare

Abstract

AI in healthcare, the word describes the use of machine learning, algorithms and software. AI has the ability to collect the data, process it and give a well-defined result to an end-user. AI does this through machine learning, deep learning and algorithms. These algorithms recognize the pattern in behaviour and create their own logic. The primary aim of the health of related AI applications is to analyse the relationships between the treatment techniques. AI-based programs are used in a diagnosis process, drug development, personalized medicine, patient monitoring and care. AI algorithms can also be used to analyse a large amount of data and then it gives the priority that which patient health report is in top and which one is on the middle and which one is at the end.

In this chapter, we have described the role and application of AI in the smart healthcare systems.

Keywords: artificial intelligence, machine learning, smart healthcare, telemedicine.

4.1 Introduction

AI is the ability of the computer to sense and analyze data provided to it. Healthcare big database [1] is useful for AI to implement machine learning algorithms. It helps in finding out necessary solutions where Humans need much more effort to do so. AI can change the Health Infrastructure by using the data provided by IoT devices over the cloud and analysing it in depth [2]. AI helps in detecting diseases, administration of chronic situations, deliver health and security services and may help in inventing new drugs or discovering drugs by using the existing database [2–4]. AI also can plan resources, depending upon the condition of the patient it can quickly allocate necessary resources (if required) for the patient and provide the information to the Doctor. In today's world, AI can be used in mobile applications, Patients can use these apps to feed data regarding their mood swings, anxiety, depression, weight height other parameters and AI could analyse this data and can provide an approximate diagnostic for the symptoms [5]. More often, smart devices track such information and upload it over the AI database. In surgeries and operations, AI can use its a database to provide the exact location of a certain wound or internal bleeding. Surgeons can use this information to carry out the operation/surgery with more accuracy. AI is also being used to detect tumours that may cause cancer in the future. We can also disburden Electronic Health Records using AI. E.H.R. developers now use AI to create better processing power and limit the time spent by the user on maintaining E.H.R. by automating the functions.

One of the most growing concerns right now is ANTIBIOTIC RESISTANCE, in countries like India the overuse of Antibiotics has caused immunity against some drugs which makes them much less effective than they once were. E.H.R. right now analyses these illnesses but by leveraging the power of AI we can develop faster solutions to prevent Antibiotic Resistance. In the field of Pathology AI can be used to pixelate images to a very high resolution [11]. What this does is enables us to get information that can easily escape the human eye. AI can detect such nuances within a matter of seconds. People who experience anxiety attacks are the first ones to call their doctors even when they experience the smallest threats. For such situations, AI enables chat boxes can be made over mobile applications or websites, which can validate the situation and provide helpful results if the problem is not too critical. AI can also create virtual Nurses that can perform a variety of tasks from talking with the patients to providing the necessary steps to be taken if the situation worsens. Some AI nurses such as Care Angel can detect symptoms using voice. Last but not least AI can be used to find genetics-Based solutions. It is estimated that Genome sequencing will become prominent in the upcoming years and might use about 200GB of data for single sequencing. AI-based optimization algorithms can optimize complex systems and the process of Genome Sequencing can also be optimized by studying and analysing data from millions of people and providing optimized solutions [12–14]. This use of AI is predicted to be the largest and most efficient use of AI ever.

4.2 Role of AI in Smart Healthcare and Other Sectors

AI refers to the simulation of human intelligence in machines that are pro-grammed to think like humans and trace their action. It gives the ability to do machines to recognize a human's face, to move and manipulate objects, to understand the voice command by humans and do other tasks. It also plays a very virtual role in the health care department. The application of AI in health care are endless. That much we know, AI has the potential to influence every step of a patient's journey from prevention and early detection, to diagnosis and treatment.

In the current generation, AI or Artificial Intelligence is one of the most popular fields in terms of research, applications, jobs, etc. It has a wide range of use in every sector and domain, which also includes healthcare as well. AI basically machines mimicking basic human cognitive functions. In laymen's terms, it is programming machines to think for themselves and make their own decisions without the help of humans. This when applied in healthcare, can become a big boost to how the whole sector in itself functions. The biggest advantage of using AI in the health sector is its gathering and processing of data and giving a well-defined result to the user/patient. These programmes can be used for diagnosis, treatments, drug development and manufacture, patient monitoring and providing personal medication. Basi-cally, every task done by a doctor can be programmed into or taught to an AI to perform it with perfection.

In Figure 4.1, we have shown various roles played by AI in various fields of healthcare. Another upcoming technology that uses bain–computer interfaces (BCI) aims to help patients with neurological diseases and trauma. Patients who are unable to communicate with their bodies and have lost the ability to move or speak can be helped through this technology. Using BCI and AI, neural activities associated with a simple body function (such as wav-ing one's hand) can be analysed, decoded and relayed to output devices that carry out the same desired functions. Such technology can drastically improve many people's lives suffering from strokes, spinalcord injuries or ALS.

Nanotechnology is already a huge part of the healthcare sector with scien-tists using this technology to target tumours inside a human body. However, operating at such atomic levels is very tricky and dangerous. This is where AI plays a very vital role. AI nanobots may currently just be a theory but their applications when manufactured are endless. These tiny 'living robots' could enter our bodies, carry out specific tasks and even help in curing cancer! AI and ML come together to form Artificial Neural Networks (ANN), which forms a new branch, called 'Deep Learning'. To give a basic idea, ANN

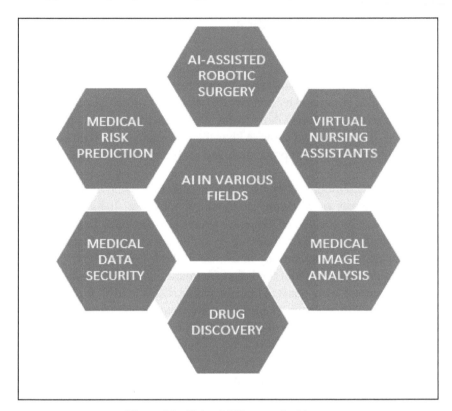

Figure 4.1   Role of AI in smart healthcare

uses the processing of the brain to model complex patterns. Through these patterns, it can then predict unseen data. In healthcare, ANN is used to predict future diseases and inform about the clinical diagnosis.

Many Radiological departments use physical tissue samples obtained via biopsies for research, but this carries considerable risks for infection. Here, AI can be used to replace these tissues with a new generation of radiology tools that are accurate and detailed. These images could also help to understand cancer cells better and aim for more appropriate treatments. With its own thinking capabilities and the information from around the world at its disposal, AI can create medicines and it even has. DSP-1181, a drug made for OCD was invented by AI through joint efforts from some companies and has even been accepted for a human trial. With the help of home monitoring equipment, healthcare professionals can keep track of patients' health conditions and provide them with the best possible treatment plan. They can also monitor their patients' actions and respond to their needs immediately. With the emergence of the IoT, health insurers can now use wearable devices to collect data about their customers and operations. This will enable them to detect fraudulent claims and make better decisions. With the iOS Health app, you can easily view all your health data in one place. It combines your data from various sources, including your iOS devices and third-party apps. Another one is the iOS Health app, which consolidates all your health data into one place [6]. It lets you keep track of all of your health metrics and see your long-term trends.

Some facts about AI are given below:

1. AI has become the new assistant of doctors.
2. AI and machine learning have the potential to transform the way we live our lives. They will also affect the way we treat our health.
3. Eric Tool stated that the healthcare industry will soon adopt AI technology.
4. The healthcare industry is expected to receive a boost from the use of AI by 2021, which is estimated to reach $6.5 billion.

The contribution of AI in smart healthcare are given below

The diagnosis of diseases is the most important area of medical education. It requires years of training and is typically not done on weekends or holidays.

1. Machine learning has made great progress in the diagnosis of diseases.
2. Machine learning is being widely used in the field of medical diagnosis. It can help in the diagnosis of various diseases and conditions by analysing vast amounts of data collected by doctors.
3. Due to the huge amount of data available in these areas and the speed at which algorithms can diagnose, they can be as successful as traditional physicians when used in specialist settings.
4. Around two billion chest X-rays are taken in a year globally. AI algorithms are more accurate than humans in analysing these examinations.
5. AI is used to gather data, decreasing the workload from humans. This is because AI makes a machine learn things by itself and perform tasks without any repeat orders. This saves time and increases work efficiency.
6. Since the number of patients keeps increasing, the data should be managed and at the same time security should also be kept in mind. In this case, if a machine works instead of humans, large amounts of data can be managed and secured with high accuracy.

7. AI can help healthcare applications by increasing the accuracy of the work. This will increase the analysis power of devices greatly. It can also be used to provide better health reports which will decrease the diagnosis time as it becomes easy to find the issue. Early diagnosis will make people even healthier and it can protect us from delayed diagnosis outcomes.

8. This use of AI in healthcare apps will not only increase efficiency but will also provide satisfactory results to everyone.

9. AI can help in the preparation of drugs. Machines make most of the drugs, but making new drugs which can be more effective than the previous updating can be a huge role played by AI. These AI-integrated machines can prepare drugs by using their own logic and ideas and prepare a drug that can provide excellent results.

10. Research will have a great boost since we do not have many researchers. AI implemented machines can accurately predict the outcomes and this will help in better diagnosis of treatments.

11. A good algorithm could increase minimum standards and reduce irrational variation by standardizing assessment and treatment based on current guidelines.

12. With AI, patients and clinicians could get local and real-time medical advice, as well as identify red flags for medical emergencies, such as sepsis, as they happen.

According to studies, doctors spend a lot of time on data entry and desk work, which leads to them not being able to interact with patients. Through automated systems, they can spend more time focusing on their patients and improving their health. In total, there are more than 2 billion chest X-rays taken in the world each year. AI algorithms are used in medical imaging systems to diagnose and detect diseases. An AI system known as Bio mind beat a team of experts from China's Brain tumour diagnosis and prognosis competition [7]. The event was held in China. It is apparent that AI has a significant impact on healthcare and since it's high on the agenda, a concern arises: what about coronavirus? Despite the fact that the virus's dissemination is still relatively new, AI-powered applications for virus diagnosis have already appeared. Infer vision, an AI company has released a coronavirus AI solution to aid front-line healthcare workers.

Following are some sectors of healthcare where AI can help:

1. Dermatology
2. Radiology

3. Screening
4. Psychiatry
5. Disease diagnosis
6. Drug creations

AI has transformed many industries around the globe [8] and it has significant utilisation in the field of healthcare. In the healthcare field, the drastic change that AI can bring are a reduction in process time. It can also resolve the data saving problem of patient's healthcare reports. It can also govern pathology labs' output by allowing for more new patients to be taken in. One case study says due to advancements in AI in this field of smart healthcare an expert pathologist's uses diagnosis during surgery is normally about 40 minutes to process, now it can take under 3 minutes only with the help of AI suggestions. AI systems have also enhanced the accuracy of analysis; this reduces time and standardizes the sample review of patients this helps in democratizing care given to patients. AI also influenced the surgical sector where we can see the AI robots are helping and performing many tasks of surgeon in operation theatre. The database used for AI should be dynamically adjustable and auto-tuned with dynamic SGA Parameters [9].

AI has reduced diagnostic error and misdiagnosis and this will be improving in treatment accuracy of a patient with more detailed results it provides that interns save direct cost due to greater precision. Although the use of AI-based technologies are having some other aspects as it reduces some employment in the field of health care. By machine learning, many scientific robots are made which help patients during the whole journey of treatment and provide a great experience to those patients and this help them to recover them faster. One of the other technical fields is Blockchain department as it assists in saving data/reports of patients in a well-defined manner. Like these field, many other technical fields also gets assisted from AI [10]. A common use of AI in healthcare involves NLP (Natural Language Processing) applications that can understand and classify a clinical document, it also helps the patient in pharma store and it gives relief over the pharma men to work in small intervals of timings as it is assisted by AI. Ruled-based expert system now uses AI in the health care field; it is widely used for clinical decision support. This leads to the help of machine learning in this field and slowly reducing/replacing rules-based systems with the approaches-based systems, on interpreting data using proprietary of medical algorithms.

We see the various uses of AI during COVID-19 pandemic. It is not sufficient enough to say that technology saves many lives but it reduces the stress from doctors who are suffering a lot during this situation. In India itself we see our doctors become warriors and act as a frontline worker. In this situation, AI played a very crucial role for both, Patients as well as Doctors too. We might think that there are great challenges to AI in healthcare, it ruins many jobs in this sector but it saves money for patients on the other hand. AI in healthcare itself performs very high levels of tasks not only in the field of medicine but also it has a huge demand over many other field/sectors too.

4.3 Use-cases of AI in Healthcare

AI increases the ability for health care professionals to better understand the day-to-day pattern and needs of the patients. So some of the main examples of the AI in the health care sector are.

1. **Keeping well:** With the rise of Internet of Medical things (IoMT) in consumer health applications. Health care applications are growing over the last few decades and it encourages positivity and a healthy environment among individuals. Now big companies as Apple have created such watches which have sensors in them and that help us to trace our heartbeat, oxygen level and distance travelled while workout and many more such amazing features are there. IoMT healthcare market will reach $137.9 billion by 2023.

2. **Diagnosis:** In 2017, Google's Deep Mind Technology helps to discover about 50+ different types of diseases only in the eye by analysing the 3D scanning. In addition, it can increase the diagnosis up to 40 times with 99% accuracy.

3. **Robotic surgery:** With the help of AI robots now healthcare industry has better surgical performance. They can easily solve the critical conditions. AI control robots have the feature that they have the 3D magnification for small cuts that increase their efficiency in cutting and stitching. In 2017 AI robots witnessed to cut a 0.03–0.03 mm the blood vessel in the Netherland.

4.4 Challenges of AI in Healthcare

AI in healthcare also faced some challenges due to the criticality of this system. It is directly connected to human life so high accuracy and precision are required for using AI-based algorithms in smart healthcare systems.

1. The biggest risk in using AI is that AI systems might be wrong at some points. Let's say in the process of detecting a tumour inside the patient,

if AI gets the wrong location, then the patient may get fooled and suffer consequences.

2. The next challenge faced by AI is the lack of accurate data. Even we are using Electronic Health Reports and continuously providing data, but the lack of authenticity persists. Even if AI learns the database but if the information provided is not relevant then there is no point in learning and storing all that data. We first need to ensure that the data, which we gather, is accurate and relevant.

3. Privacy plays a major role in our lives. We need to develop a system that we can trust and ensure that our data is not leaked.

4. AI must gather relevant data throughout the world. If only a part of the population is used to analyse the data, then the remainder, which might have different traits or diseases, may not be accounted for future use.

5. Tools could be mistaken with confidence and algorithms could be misleading. Unsafe AI could harm healthcare providers and patients.

6. Some surgeries are too costly.

7. Due to overdose of some rays can also cause skin or body cells damage.

8. Sometimes the result produced after check-ups have blunder mistakes.

4.5 Conclusion

The willingness of people all across the world to use AI and robotics is growing. We may conclude that the need for speedier, more intuitive and low-cost health services is the key driver of this increase. Increased usage and acceptance of technology requires trust; however, 'human relations' remain an important part of the healthcare experience. Therefore, it appears that we will be able to obtain.

References

[1] P.K.D. Pramanik, S. Pal, and M. Mukhopadhyay. Healthcare big data: A comprehensive overview, in: N. Bouchemal.

[2] P. Ahlawat, and S.S. Biswas, (2020). 'Sensors based smart healthcare framework using internet of things (IoT)', International Journal of Scientific and Technology Research, 9(2), pp. 1228–1234.

[3] B.A. Thakkar, M.I. Hasan, and M.A. Desai, (2020). 'Health care decision support system for swine flu prediction using naïve bayes classifier', *International Conference on Ad-vances in Recent Technologies*

in Communication and Computing, Kottayam, pp. 101–105, doi: 10.1109/ARTCom. 2010.98.

[4] S. Aditya and B. Ranjit, (2013). An algorithmic approach for auto-selection of resources to self-tune the database , IJRIT International *Journal of Research in Information Technology,* 1(9).

[5] V. Raju, et al., (2020). Artificial Intelligence (AI) applications for CORONA VIRUS DISEASE pandemic, Diabetes metabolic syndrome: *Clinical Research Reviews,* 14(4), P. 337–339.

[6] A. Bhushan, et al., (2017). 'I/O and memory management: Two keys for tuning RDBMS', Proceedings on 2016 2nd *International Conference on Next Generation Computing Technologies,* NGCT 2016, 7877416, pp. 208–214.

[7] I. Khanchi, and E. Ahmed, (2019). 'Automated framework for real-time sentiment analysis', *International Conference on Next Generation Computing Technologies,* NGCT.

[8] T. Singh, et al., (2017), Detecting hate speech and insults on social commentary using nlp and machine learning. *International Journal of Engineering Technology Science and Research* 4(12), 279–285.

[9] S. Kumar, S. Dubey, and P. Gupta, (2015). 'Auto-selection and management of dynamic SGA parameters in RDBMS', 2nd *International Conference on Computing for Sustainable Global Development* (INDI-ACom), 1763–1768.

[10] J.C. Patni, et al., 'Air quality prediction using artificial neural networks', *International Conference on Automation, Computational and Technology Management* (ICACTM).

[11] P. Koranga, et al., (2018). Image denoising based on wavelet transform using visu thresholding technique. *International Journal of Mathematical, Engineering and Management Sciences,* 3(4), 444–449.

[12] G. Negi, et al., (2021). Optimization of complex system reliability using hybrid grey wolf optimizer. *Decision Making: Applications in Management and Engineering,* 4(2), 241–256.

[13] A. Kumar, et al., (2018). Complex system reliability analysis and optimization. *Advanced Mathematical Techniques in Science and Engineering,* 1, 185–198, River Publishers.

[14] A. Kumar, et al., (2019). Multi-objective grey wolf optimizer approach to the reliability-cost optimization of life support system in space capsule. *International Journal of System Assurance Engineering and Management,* Springer, 10(2), 276–284 https://doi.org/10.1007/s131 98-019-00781-1.

5

Application of IoT in Smart Healthcare

Abstract

IoT is a platform where the patient can meet the doctor and show their report in a virtual manner and take the proper guidance without physically meeting. A few years ago, diagnosis of diseases in the human body was only possible when the patient is physically meeting with the doctor and came to the hospital for check-ups/treatment. This in turn increases the healthcare cost and stained the healthcare facility in rural and remote locations. The technological advancement through these years allowed us to diagnose of various diseases and health monitoring. In the last decade, there are so many technological developments are there with the help of these technologies. Now one can monitor the patient blood pressure, level of glucose, oxygen level and so on in an easier way and the best thing is there is no need to physically meet with doctors to show reports.

In this chapter, we have described the role and application of IoT in the smart healthcare systems.

Keywords: Internet of Things (IoT), sensors, telemedicine, smart healthcare.

5.1 Introduction

The Internet of People was invented in 1983 and used to share information such as pictures, texts and audio files between human beings. IoT is the same only the difference being that, in this case, things have specific sensors which can interact with other things and human beings. Times like Covid-19 pandemic have forced us to find an alternative way to visit doctors and get the necessary consultation. IoT provides an excellent method to solve this problem for both patients and doctors. For patients, everything they use or wear such as smartwatches, weighing machines, mobile phones and blood-pressure machines can be embedded with necessary sensors, which provides constant feedback of the person and upload/analyse it over the cloud. For instance, a smartwatch with having a spO2 calculator can notice a sudden drop in the oxygen level of the person and can send the feedback to the nearest emergency service manager. This could initiate a call for an ambulance and the person can get immediate assistance. From the point of view of doctors, they could use smart machines, which provide them with data regarding the most common illness in an area and then this data can be analysed by the doctors and scientists to determine the most common cause of that illness. This approach can help in eradicating some diseases. With increasing technologies, we can connect a wide range of devices ranging from ordinary everyday use devices to very sophisticated industrial devices through the internet using certain devices and make its functioning easier. Moreover, this had been possible with the help of the IoT. It enables us to create a network between physical objects or things that are embedded with various technologies or sensors. Its purpose is to exchange information with other devices over the internet [1].

In Figure 5.1, we have shown the various uses of IoT in smart healthcare.

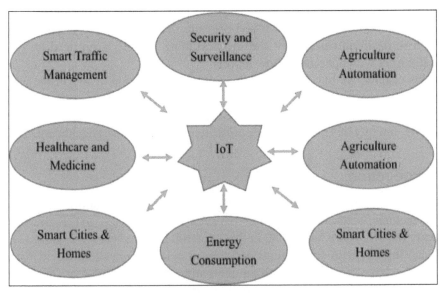

Figure 5.1   IoT usage in various sectors

IoT has many industrial benefits: such as in automotive, manufacturing, transportation, retail, healthcare, public sector and many more. With the increasing global population and depleting global health, a large section of society is prone to various chronic diseases [2, 3]. But not everyone has access to healthcare services. An example can be taken of the recent pandemic situation where due to large various reasons people were unable to access basic healthcare services but this is not the end. The development of IoT in the healthcare industry has enabled to cure this situation use to great extent. Though IoT will not eradicate the disease but it will at least lessen the need for hospitalization by providing treatment and improved lifestyle.

The Hospitals, which use machines such as Glucometers, ECG machines, inhalers and pacemakers, can be embedded with such sensors. This will lead to improved communication between the machines so that the dose of some medicines can be self-adjusted by these devices with the data provided to them. Another breath-taking approach is the use of 'eatable/ingestible sensors', procedures such as a colonoscopy, which demands the presence of a sensor inside the body that can be covered by this single device. They can also help in pinpointing in case the stomach becomes too alkaline/acidic or if there is internal bleeding. Usually, such insights are impossible without visiting the doctor, but the IoT has made it possible to diagnose such conditions virtually and provide the necessary treatment.

5.2 Role of IoT in Healthcare

IoT refers to the physical objects or devices which are inter-connected through the internet and all collectively taking in and sharing data without requiring human touch or interaction. In the last few decades, IoT has spread in every household, connecting our phones, kitchen appliances and other electrical devices (Figures 5.2 and 5.3).

Similar to our household, IoT also plays a huge part in healthcare. Cases for the same are given below:

1. In a hospital, various devices are used to track and monitor patient's heartbeats, their psychological condition or their movements. All such instruments are connected to each other and provide real-time information to the doctors or nurses. Such technology helps the staff to keep a check on all the patients at the same time and even reduces hospital stay time [4].

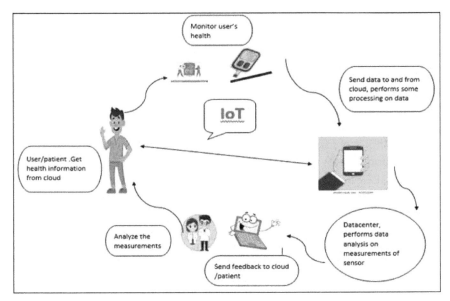

Figure 5.2   Data flow in smart healthcare application using IoT

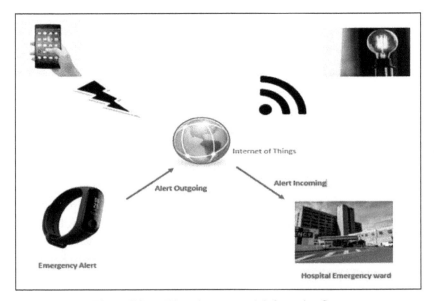

Figure 5.3   IoT-based smart watch information flow.

2. IoT also has a huge role in our daily lives. Everyday use devices, such as fitness bands, oximeters, sphygmomanometers, etc. can be tuned/programmed to provide daily reports on calories, blood pressure, oxygen levels and even help in booking appointments. IoT can also help in preventing emergencies by sending alert signals to close kin and concerned authorities when there is a disturbance in daily patterns or reports of the person.

3. Another huge part played by IoT in healthcare is in Healthcare Insurance companies. Such companies make use of large amounts of data and information collected extensively from various healthcare facilities and health-monitoring devices. This data helps them in understanding the most common ailments, injuries and diseases and coming up with the most helpful and profitable insurance policies for their customers. It also aids in bringing transparency between people and themselves in terms of prices, terms and conditions etc.

Some examples of IoT-based smart healthcare devices are.

1. Remote patient monitoring
2. Glucose monitoring
3. Heart-rate monitoring
4. Depression and mood monitoring
5. Hand hygiene monitoring

5.3 Users of IoT in Healthcare

IoT has helped the healthcare sector effectively by decreasing the workload and increasing the efficiency in problem-solving. IoT devices help to continuously monitor a person and provide real-time results. This saves the time of the doctor as well as the patient. The data captured by IoT devices will be stored in cloud storage and analysed by AI-enabled digital interfaces systems for prediction [5, 6].

IoT describes the use of physical devices and connectivity to enable the exchange of data in the healthcare sector. It is a new and upgraded technology, unleashing the potential to keep patients safe and healthy and permitting doctors to deliver excellent care. Ageing or exterminating chronic diseases cannot be stopped neither can technology do so but at least technology can make healthcare more easy and accessible. This technology-based healthcare method has been proved beneficial to all the healthcare sectors who were looking forward to an easy and feasible method related to healthcare solutions. Mentioned below are a few ideas on how Iota applications are used in the healthcare sector.

1. **Remote patients monitoring:** The IoT assures ever-greater levels of interconnectedness and data sharing, enabling connected devices to connect virtually anywhere. Healthcare is one area where the IoT thrives, as remote patient monitoring and healthcare IoT emerge almost seamlessly together. With IoT devices like glucose meters, fitness bands, etc., health metric information is collected automatically from patients without having them physically present at the hospital or clinic. The devices allow the information to be collected without patients having to travel to the hospital or clinic to collect it. IoT devices collect patient data, which is then forwarded to healthcare professionals or patients that can view it through the software application. The data is analyzed using various algorithms in order to create alerts or make treatment recommendations.

2. **Glucose monitoring:** According to WHO, the total number of diabetics in the world is 422 million in 2014. A growing number of healthcare companies have developed monitoring equipment for diabetics to support a functioning lifestyle. Using IoT systems, glucose-monitoring systems have been introduced in the last few years to improve the quality of health care. Sensors and computer systems are designed together to provide real-time glucose, body temperature and context data on a graphical and human-readable basis to end-users, such as patients and doctors. Monitoring of glucose levels can be done continuously in real-time with this system.

In the medical field IOT is said as I.O.M.T. here, I.O.M.T stands for Internet of Medical Things. IOT has provided many applications, gadgets in the field of the medical industry that is benefited to patients, families, physicians, hospitals and also to insurance companies as well. In the case of patients, I.O.M.T has provided many wearable gadgets like fitness bands, active smart garments and many other wirelessly devices which are used to measure things like blood pressure and heart rate monitoring cuffs, glucometer, etc. This also gives a great outcomes and many lives can be saved with help of I.O.M.T devices. In the Case of hospitals/transports, due to the introduction of IOT techniques, the billing is also going fast and reception employment is also available at ease. Now, such a large gathering near the counter is vanished in case of transport (ambulance facilities) as well due to the use of I.O.M.T. Machine Learning based algorithms are also used in I.O.M.T. for analysing medical images [6]. This is also useful for Telemedicine systems [7].

Nowadays in pandemic of COVID we all see how drastic change came to seen in the field of health care as we saw there are a scarcity of beds so doctors are started using IOT devices and software to interact with their patient who was not able to visit them on that situations, names of such software as

1. Doctor 24x7
2. Practo
3. DocsApp
4. Curofy
5. 1mg
6. Ask Apollo, etc. i.e., there are many more such platforms was a patient directly contact his/her doctor. This is all because of the advancement of technology and introduced of the IoT [8].

Hence, we say that IoT has a great impact on Health care and this gives ease to live.

Some of the benefits of IoT in healthcare are given below

1. For monitoring patients with the minor diseases which will use fewer hospital resources.
2. To provide healthcare facilities to people living in rural areas.
3. Enable elderly people to live independently at home and take care of their health.
4. Real-time monitoring through IoT platforms can save a life during emergencies [9].
5. To collect and save data and reduce the need for manual work to a great extends and provide auto-detection feature [10].

5.4 Challenges of IoT in Healthcare

The biggest challenge faced by IoT is security. Many multinational corporations may steal the data provided by the IoT devices and it could be sold to fraudulent parties worth millions of data. This problem is so huge that there have been numerous cases of hackers stealing information via these devices and scamming the common people. There should be a way to prevent data leakage and modification via hackers. The next challenge is to implement IoT not only in urban settings but in rural areas as well. Devices, which are optimized with IoT usually, cost more and it becomes difficult for the common man to get access to those devices. Lastly, we need to cultivate the ideology of believing in IoT among young people so that they get interested in such ideas and provide solutions that are viable and feasible. Blockchain is providing a major solution for Security [11]. Effective integration of Blockchain can help to overcome one of the challenges of IoT in healthcare [12].

5.5 Conclusion

IoT refers to a network of physical objects, which can gather and share electronic information. The IoT includes a variety of smart devices, ranging from industrial machines to sensor that tracks the human body. The key concept of IoT is that it facilitates us according to our comfort. In addition, by 2022 the market value of IoT will be about $9 trillion. IoT also enables to save of the data of the patients on a large scale. As the data is the goldmine for the healthcare stakeholders to improve patient's health and experiences. IoT Enable patient monitoring only when needed hence reducing the unnecessary visit to the doctors. With the help of continuous monitoring, disease can be diagnosed easily and accurately and Helps physicians to provide suitable medicine according to the symptoms. Therefore, IoT helps a patient and doctors at each step in the health treatment of a patient.

References

[1] P. Ahlawat, and S.S. Biswas, (2020). 'Sensors based smart healthcare framework using internet of things (IoT)', *International Journal of Scientific and Technology Research* 9(2), pp. 1228–1234.

[2] S. Taneja, and E. Ahmed, (2019). 'I-Doctor: An IoT based self patient's health monitoring system', *International Conference on Innovative Sustainable Computational Technologies*, CISCT 2019.

[3] S. Kumar, et al., (2015). 'Auto-selection and management of dynamic SGA parameters in RDBMS', *2nd International Conference on Computing for Sustainable Global Development* (INDIACom), 1763–1768.

[4] J.C. Patni et al., 'Air quality prediction using artificial neural networks', *International Conference on Automation, Computational and Technology Management* (ICACTM).

[5] N. Kamdar, et al., (2020). Telemedicine: A digital interface for perioperative anesthetic care, Anesthesia and Analgesia, 130(2), p. 272–275, doi: 10.1213/ANE.0000000000004513.

[6] K. Kshitiz and Shailendra, (2018). NLP and machine learning techniques for detecting insulting comments on social networking platforms.

International Conference on Advances in Computing and Communication Engineering (ICACCE), IEEE, p. 265–272.

[7] V. Lubkina, and G. Marzano, (2015). Building social telerehabilitation services, *Procedia Computer Science* 77, 80e84.

[8] H.K. Sharma, et al., (2019). 'Real time activity logger: A user activity detection system', *International Journal of Engineering and Advanced Technology* 9(1), pp. 1991–1994.

[9] E.A. Khanchi, (2019). 'Automated framework for real-time sentiment analysis', *International Conference on Next Generation Computing Technologies* (NGCT-2019).

[10] T. Singh, et al., (2017). Detecting hate speech and insults on social commentary using nlp and machine learning. *International Journal of Engineering Technology Science and Research* 4(12), 279–285.

[11] A. Saha, et al., (2019). Review on 'blockchain technology based medical healthcare system with privacy issues'. *Security and Privacy*. 2. 10.1002/spy2.83.

[12] S. Salman, et al., (2020). A secure blockchain-based e-health records storage and sharing scheme, *Journal of Information Security and Applications*, 55, 102590.

[13] H.M. Hussien, et al., (2019). A systematic review for enabling of develop a blockchain technology in healthcare application: Taxonomy, substantially analysis, motivations, challenges, recommendations and future direction. *Journal of Medical Systems*, 43(10). doi: 10.1007/s1 0916-019-1445-8.

6

Application of Blockchain
in Smart Healthcare

Abstract

Blockchain is a distributed ledger technology with IOT, which helps to make machine-to-machine interaction possible. So, we can also say that it is a system that records information in such a way that makes it difficult to change, cheat or hack. Smart healthcare needs blockchain technology to store the digital records of patient health. Hence to make healthcare system more secure and robust against tempering and theft of electronic health records (EHR).

Keywords: blockchain, smart contract, smart healthcare telemedicine.

6.1 Introduction

Healthcare accounts up a significant portion of the gross domestic product (GDP) in developed countries. Hospital expenditures, on the other hand continue to rise, due to wasteful procedures and health data breaches. This is the area, where blockchain technology has the potential to improve things [1]. It is capable of a wide range of tasks, including secure encryption of patient data and the management of epidemics. Estonia is a pioneer in this arena, having implemented blockchain technology in healthcare in 2012. When data is transmitted from the smart device to the doctors or hospitals it is very much possible that a hacker can manipulate the data being sent. Such cases in the history have caused to loss of millions of dollars and unfounded patient history. The answer to this problem is the use of blockchain technology. Blockchain technology is based on the concept that there is no central authority which regulates and checks the transactions that are taking place [2]. A Public ledger is made available over which all transactions are recorded and can be seen by doctors/health professionals only. Whenever a transaction takes place, it uses a cryptographic hash function such as the SHA-256 to authenticate the transaction. As it used a hash function the hash key is only present between the sender and receiver, the only task left to do is used supercomputers to match the hash keys and authenticate the transaction. Some European countries such as Estonia are using only blockchain technology to maintain an excellent electronic health record (E.H.R.) of its citizens and storing it for faster access of information. Estonia uses keyless signature infrastructure (K.S.I.) to perform the task. Blockchain is a new technology that comes in the market recently and this technology promises an efficient, cost-effective, reliable and secure system for do any type of recording, transaction without the need of middleman [3]. Blockchain database is a distributed database that records the data for that time and provide a link between the user and the organization where the transaction should be done. Due to this technology, no one can alter any transaction data. Member of blockchain who do a transaction is called nodes [4]. Blockchain technology provides various types of nodes to enter in the blockchain network.

In Figure 6.1 we have shown the characteristics of blockchain. With the help of blockchain we can introduce these characteristics in healthcare also [5].

The main features which make blockchain the backbone of secured and seamless transactions is:

1. Ability to have a decentralized and transparent log.
2. Establishing private connections.
3. Protecting the identity of any individual with complex and secured codes.
4. Making information immutable.
5. Irreversible transactions.

Blockchains are immutable databases, which record information and data. What makes Blockchain different is that it is impossible to tamper, hack or change the information stored inside it. Most famous use of blockchain is in crypto currencies such as Bitcoin, Dogecoin etc [6]. Similar to its use in crypto currencies, blockchains also have tremendous potential in healthcare. Some popular uses for the same are management of electrical medical records (EMR) and protection of healthcare data. Communication between staff costs a huge amount of money to healthcare sector. Pair that with the time

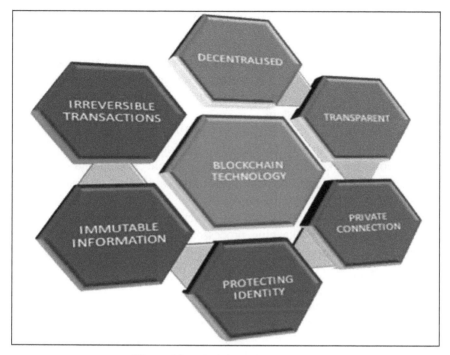

Figure 6.1 Blockchain characteristics

consuming process of finding the patient's records, there is a delay in patient care as well as exhaustion of the staff and their resources.

However, these problems can be solved easily with the help of blockchains. Blockchains creates a single pit stop for all the patient data, which can be quickly and efficiently accessed by the doctors and nurses. All this data collected through treatments of various patients can be saved in the EMR. These massive amounts of data can then be accessed by various companies for research purposes to create or find new cures for diseases [7]. This further boosts/promotes clinical research and helps in growth of the healthcare sector. Counterfeit medicines are a huge cause of death annually. So, ensuring authenticity and making sure medicines arrive from the correct place is a big challenge. Blockchains help cancel out such problems by utilising proper system to track items from manufacturer to buyer. And since blockchains can neither be hacked nor tampered with, the data remains absolutely safe.

6.2 Application of Blockchain in Healthcare

Basically, blockchain is kind of tool in Database. Which allow to stores data over the cloud or internet, also provide permission to use and also secure these data as a trust. There is a very important note that we cannot edit those data present in blockchain. Following figure (Figure 6.2) represents the flow of information in blockchain system.

Blockchain technology firstly introduced in 1991 by Staurt Haber and W. Scott Stornetta. But this technology got the identity in 2009 by the launch of Bitcoin. Blockchain is called because it collects data and information in form of block and it contain all information in form of hash and each block contains information of previous block and this forms a chain like structure so it is called blockchain [8]. Blockchain technology has the potential to impact all recordkeeping processes, including the way transactions are initiated, recorded, authorized, and reported. Blockchain could be used to securely and efficiently transfer user data across platforms and systems

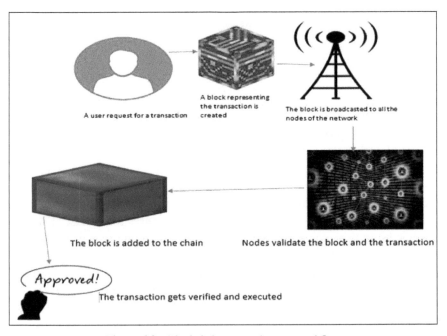

Figure 6.2　Blockchain transaction approval flow

Blockchain basically works on three fundamentals:

- Incentives
- Identity
- Trust

Nowadays as technologies are getting enhanced, this brings a drastic change in many other fields with the help of blockchain. One of such field is healthcare, it provides a table for secure a patient's useful information such as reports of blood pressure, sugar level, ECG reports, and many other reports of patients are secure and saved in cloud form over the internet and this makes an ease for user [8]. Since, in early days we see patients are worry for their old reports or they were lost. Doctors also face some issue during treatment of such patients as loss of information. In following figure (Figure 6.3), various users of smart healthcare who will use Blockchain is shown.

Blockchain technology has the potential to impact all recordkeeping processes, including the way transactions are initiated, recorded, authorized, and reported. Blockchain could be used to securely and efficiently transfer user data across platforms and systems. All-important loss of data and information of any such problems are overcome after introduced of blockchain technology. Now the patient's problem look like move very far away from them as those important reports are stored over the blockchain though which they can easily access on time of use or in any emergency. Doctors have also got an ease in check-up of any patients with the help of blockchain data as

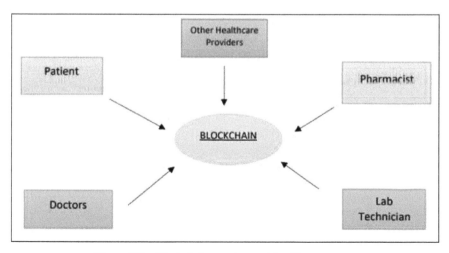

Figure 6.3 Blockchain user in smart healthcare system

that very patient is either suffering from earlier from any such problems which may leads to this new kind of problem to that very patient [9].

Yes, we can say the blockchain technology bring significance change over worldwide as well this shows new era of development in E-Medical field. In past one year, these technologies give new view or image in medical field. As in pandemic of COVID we had seen, doctors and patients interaction through virtual mode and the medical data of patient are easily available to doctors whenever they needed it. This also secure many of lives of patients. blockchain is set up of records in an exceedingly system that makes it laborious to corrupt the information. Stuart Fritz Haber introduced blockchain technology and W. Scott Stornetta in 1991. They wished to determine a system wherever document time-stamps couldn't be tampered with. Blockchain could be a sort of info. It stores knowledge in an encrypted manner then ties them along. Most of the developed countries pay an honest quantitative relation of gross domestic production health care sector. However, rising in value of hospitals and threats of health knowledge breaches still continues. This is where; the blockchain technology plays its significant role. It will multitask between encrypting patient data and handling epidemics. E-health records enable automatic change and sharing of medical data on a given patient among a corporation or network of organizations. This allows all the info to be accessed from one place with more help analysers in clinical research, safety event and adverse event reportage and identification, and public health reportage.

In Figure 6.4, the integration of various blocks in a blockchain is shown.

By avoiding miscommunication between doctors caring for same patient, numerous mistakes can be avoided. Quicker designation and interventions will become attainable and care will be personalized to every patient. The secure options related to the blockchain will facilitate and shield health

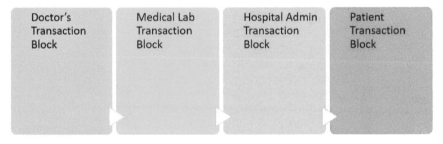

Figure 6.4 Connection of blocks in blockchain

records far well. Every individual contains a public symbol or key and a personal key, which might be unlatched solely as and for the amount necessary.

- **Increased privacy and security:** As we have discussed earlier, privacy is the main purpose of using blockchain. This means all the medical records of an individual are stored safely. Efficiency of blockchain is one of the main reasons for the growth of it in healthcare. All the records are encrypted so only a person who has required information to access it can access it. Therefore, a person need not to worry about leaking of data.
- **Readily available:** As the industry is growing rapidly, the use of these blockchain techs have also increased. This provides increased accessibility for the people.
- **Tracking the supplies:** Besides privacy and security, blockchain tech can also be used to keep an eye on the supplies. For example if we have ordered something from a medical provider, the customer can track it. MediLedger is one of the examples that uses blockchain to increase the authenticity and transparency of the goods/supplies.
- **Verification of the team:** Blockchain can also be used to verify the credentials of the health support. This makes an individual know more about his team. This can lead to better relationships with the team. ProCredEX, an American company, has built an application, which verifies the details of the support.
- **A focus on transactions:** The validity of transactions will be determined by consensus, and while transactions can be interpreted widely though not all healthcare transactions will be appropriate for this technology. With blockchain, smart contracts are possible, which are downloadable computer code that allows programs, functions, or transactions to be automated. In addition, the technology provides a way to create value for transactions through tokens. Also, crypto currency settings have been implemented with tokenization.

6.3 Challenges of Blockchain in Healthcare

The major challenges in using blockchain technology for health reformation are scalability and settlement time. The blockchain usually scales at O(n) this means the size of chain grows directly proportional to the number of elements which are added to it. Settlement time of blockchain due to the presence of very large number of people is quite high and averages to about 5–15 minutes per transaction. This is one of the major drawbacks on as to why blockchains are not used. The last but not the least is the money factor. All cryptographic hash functions require very large number of calculations which is feasible by expensive equipment only.

Although DLT can be used in a variety of healthcare settings, not all activity in healthcare is tied to transactions. However, because the data in public, blockchains is readily available, they cannot be utilised to store private information such as identifiable health data. As a result of this transparency, providers are required to examine privacy problems in order to ensure that protected health information is maintained (PHI). Second, while blockchain technology is vulnerable to some assaults, it also provides built-in defence against others. The blockchain's code makes it vulnerable to zero-day attacks, vulnerabilities, and social engineering. As a result, information security must be given special consideration, especially when it is employed in healthcare.

Blockchain is basically a chain of transactions which is used to share records. They are mainly used when an organisation or system needs to save secured data. It provides network security from basic to high level. We can also say that it provides an authentication to all the users. We can also say that it is used to manage electronic medical record data, personal health record data. Besides the secure transfer of patient record, it can also manage medicine supply chain. The biggest advantage of using blockchain in real time applications is that the data remains secured and can only be shared with the respective patients. In many areas, it can prevent miscommunication between doctors and the patients medical record. As the doctors has the access to the medical records which creates one ecosystem where the data may include doctors, hospitals, prescriptions and treatment involves at any stage in life. This technology of blockchain will lead to faster prognosis. The researchers can also use this as a database for their researches. Insurance company to claim health insurances can also use it. As seen in the flowchart we can say that if a person had an accident. He would be unconscious. While getting checked doctor will ask a few questions. Being a little unconscious what if the diabetic person could not respond [10]. Blockchain network can allow patients to append or change all the information that can be then seen by the medical staff. Therefore, through these networks of blockchain as a patient we can provide the access for our information and it will always be updated.

Some of the major challenges for blockchain in healthcare are given below.

- **A single source of truth:** The technology offers the primary benefit of providing a single point of truth for those who participate. This is because consensus must be reached by all nodes. So participants are bound by the same understanding of the nature of the data in the network.

- **Lack of privacy and security vulnerabilities:** Public blockchains, especially those that are available to the public, do not provide the ideal storage environment for private data. blockchains in healthcare offer significant privacy benefits for protected health information (PHI) when used for certain transactions.

6.4 Conclusion

In the healthcare industry, blockchain has a wide range of applications and uses. Aside from facilitating secure patient medical record transfers, ledger technology also assists researchers in unlocking gene sequences. Though it utilizes well-known and tested underlying technologies, such as network connectivity, hashes, and encryption, this technology is unique from traditional programming, networks, databases, and web interfaces.

References

[1] A. Saha, et al., (2019). Review on "Blockchain technology based medical healthcare system with privacy issues". Security and Privacy. 2. 10.1002/spy2.83.

[2] N. Kamdar, et al., (2020). Telemedicine: A Digital Interface for Perioperative Anesthetic Care, Anesthesia and Analgesia, 130(2) p. 272–275, doi: 10.1213/ANE.0000000000004513.

[3] S.T. Wu, and B.C. Chieu, (2003). 'A user friendly remote authentication scheme with smart cards', Computers Security, 22(6), pp. 547–550.

[4] H.M. Sun, (2000). 'An efficient remote use authentication scheme using smart cards', IEEE Transactions on Consumer Electronics, 46(4), pp. 958–961.

[5] S. Salman, et al., (2020). A secure blockchain-based e-health records storage and sharing scheme, Journal of Information Security and Applications, 55, 102590.

[6] K. Tanesh, et al., (2019). Kumar and An Braeken and Anca Delia Jurcut and Madhusanka Liyanage and Mika Ylianttila, "AGE: authentication in gadget-free healthcare environments", The Author(s).

[7] M. Zahid, et al., "Authentication and Secure Key Management in E-Health Services: A Robust and Effi-cient Protocol Using Biometrics", Corresponding author: Anwar Ghani.

[8] C. Preeti, et al., (2019). "Cloud-based authenticated protocol for health-care monitoring system", Springer-Verlag GmbH Germany, part of Springer Nature.

[9] R. Eswaraiah, et al., (2020). Rayachoti1 Sudhir Tirumalasetty1 and Silpa Chaitanya Prathipati, "SLT based watermarking system for secure telemedicine", Springer Science+Business Media, LLC, part of Springer Nature.

[10] Gupta, et al., (2021). "User Anonymity based Secure Authentication Protocol for Telemedical Server Systems", Int. J. of Information and Computer Security.

7

Security and Privacy challenge in Smart Healthcare and Telemedicine systems

Abstract

Smart healthcare and telemedicine Systems are available for usage by patients and doctors. Many industries are providing mobile/web based e-healthcare and TMIS systems where patients and doctors can interact with each other using remote services. These systems are using cloud storage to store patients health related digital records. Security and privacy maintenance is a challenge in these cloud based services. In-efficient authentication schemes for authenticating users (patients, doctors, medical staff etc.) is also a challenge in widely usage of these systems. Digital health record is a critical information for each patients. If a hacker hacks these medical reports and make some tempering in these documents by changing their prescription given by the doctors then it will affect the health of the patient. In this chapter, we will discuss the current challenges in security and privacy in using smart healthcare systems.

Keywords: smart healthcare, e-hospital, telemedicine, cloud storage, privacy and security.

7.1 Introduction

Smart healthcare and telemedicine systems are works on network. So Security and privacy management is one of the major challenge in using smart healthcare systems. There are many factors where the security and privacy will be compromise in these systems. Network attackers can introduce many threats in these systems if not properly managed by the providers. These threats may cause serious health issues for patients.

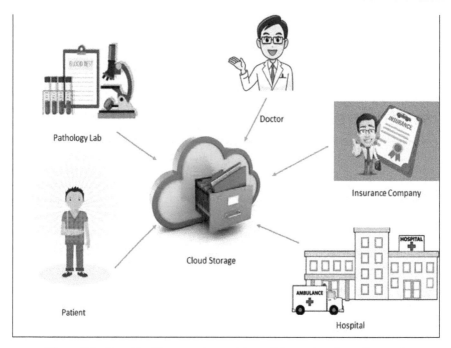

Figure 7.1   Smart healthcare system or telemedicine system (user diagrams)

We know that there are many users of a smart healthcare and telemedicine systems. In Figure 7.1 we have shown a basic user connectivity and accessibility diagram of a smart healthcare system.

As we can see in the above diagram, there can be various kind of users for a smart healthcare systems which includes:

i. **A patient:** A user which requires diagnosis of his/her health issue and need prescription for the same.

ii. **A doctor:** A user which provides diagnosis of patient health issue and gives prescription for the same.

iii. **A hospital:** A place where a patient get admission for treatment of health disease. It stores patent records in its local storage.

iv. **A pathology lab:** A place where a patient gives his body samples for testing purpose for diagnosing the disease on recommendation of the doctor.

v. **An insurance company agent:** An agency, which provide financial support, bears the cost on behalf of insured patient and pay hospital bills after getting required information for smart healthcare cloud storage.

Table 7.1 Security services required in smart healthcare

Security Service	Description
Confidentiality	TMIS system should be accessible by the authenticated users. Information should be encrypted. Individual user should access to his/her records only.
Availability	It should be available 24X7. It should be accessible from anywhere. It should be fast performing. Stored HER should be easily available for Authenticated user.
Integrity	EHR records or other critical information should be same at all servers. It should be digital format. It should be encrypted so that unwanted changes can be avoided.

There may be some other users of these smart healthcare system. We have mentioned only the important and frequent users for these systems. In Table 7.1, we have listed some security services required for a smart healthcare system. There services need to be provided by TMIS systems. These security services are mandatory to be taken care by TMIS systems for the patients and doctors.

7.2 Literature Review

Authentication algorithm for accessing remote system introduced in 1981. A new authorization advanced scheme based on smart cards was proposed for user authentication purpose with more benefits over previous authentication system. They proposed more user-friendly authentication model based on smart card for remote users [1]. Wu and Chieu resolved the shortcoming of Sun (2000) smart card based model in their model. Lee and Chiu reviewed the Mu and Chieu Model and identified a forgery attack in their authentication system and they proposed an improved model after fixing all security flaws and possible attacks. An advanced scheme came up with a one-factor authentication system using the Diffie Hellman key exchange theory and they used one-way hash function in to secure unsafe networks and provide security for some general network attacks like stolen-verifier attack, replay attack, guessing attack and modification attack. In 2011 an author declared an secure and efficient authentication system with some unique features. It includes no need to maintain password documentation, no need to update master key every time a new service provider joins to the system.

Wu and Chew introduced a three-factor authentication system for TMIS using biometric system and random numbers [2]. Analysing their schemes, Tan in 2014 caught an issue of user anonymity and a reflection attack in their work. Arshad and Nikooghadam in 2014 identified replay attacks and denial-of-service in Tan's scheme. Sun reviewed Lee and Liu's (2013) user authentication scheme for TMIS and found an offline password guessing attack vulnerability. In same year, Amin and Biswas (2015) identified some flaws in authentication schemes of Giri et al.'s (2015). These flaws or weaknesses are: offline password guessing, lack of user anonymity and insider attacks and proposed a new RSA-based scheme.

In continuous effort, to enhance the security in previous schemes, Bin Muhaya in 2014 introduced an enhanced authentication and key agreement scheme for TMISs. However, it was proved that Bin Muhaya's scheme was also has a weakness for off-line password guessing attacks and it is not able to perfect forward secrecy. To remove the security weaknesses of authentication scheme of Bin Muhaya's, Shamshad et al. proposed a new two-factor key agreement scheme and user anonymity preserving authentication for TMISs [4].

7.3 Security Challenge in Smart Healthcare

Patient medical records are stored digitally as EHRs for easy availability but there are some technical challenges and vulnerabilities to storage and access of medical records in cloud databases [5, 6]. Another problem is sharing of medical data with others and at the same time ensuring privacy and integrity of data. At the same time, patient does not have complete ownership over access of their medical records [7, 8]. Even if access control is applied there are chances of data leak if an adversary gets access to the database in which EHRs are stored, hence the data should be stored in encrypted form. A proper history of patient record access and update should be maintained and patient's permission should be required while editing his/her records to ensure that data is not edited by an unauthorized person or by an adversary [9]. There are various security threats in smart healthcare systems.

In Figure 7.2, we have shown some major security threats that the users of these systems may face.

There are various networks attacks and threats but we have mentioned only those are more related to these systems.

In this diagram, we have presented six major security and privacy threats to these systems.

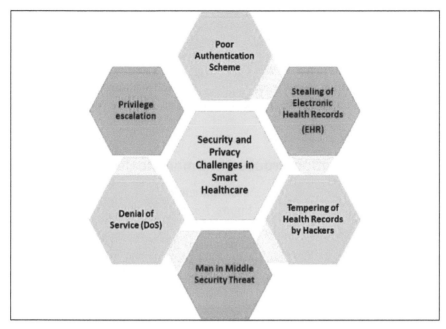

Figure 7.2 Security challenges in smart healthcare

i. **Poor authentication scheme:** This is a challenge where a non-authorize user can access these system and create some unwanted modification or deletion in heath related data.

ii. **Stealing of electronic health records (EHR):** This is a challenge where a user can steal the EHR, temper them, due to which a major problem may occur for patient [10]. So original information will not be available and patient will not be able to get right prescription.

iii. **Tempering of health records by hackers:** This is a challenges where a user can temper a patient EHR due to which a major problem may occur for patient. Because original information will be modified which may cause for improper prescription.

iv. **Man in middle security threat:** In this security threat, a third person can take control of network communication between patient and doctor and behave like a doctor for a patient and behave like a patient for a doctor and communicate with any of them as real patient or doctor.

v. **Denial of service (DoS):** This is a major security challenge. In this issue a doctor or a patient can deny for giving prescription or receiving prescription. A doctor may deny in this issue that he has not given a certain prescription to the patient and in same way a patient can also deny that he has not taken any prescription from any doctor [11].

vi. **Privilege escalation:** In this kind of security threat, a network attacker can change the privileges of real users of these systems. Doctor privileges can be transferred to the patient due to which without any medical knowledge a patient can prescribe to other patient, which is very harmful for real patient.

7.4 User Authentication Process in Smart Healthcare

In this section, we are explaining a user authentication process in smart healthcare or telemedicine system. In Figure 7.3, we have shown a two-way authentication scheme. In this scheme a user need to register first to a smart healthcare system using a username and password than system will generate a smart card for the user, which will be required at the time of login [12].

It is considered as two-way authentication scheme as given below.

i. First way is user name and password.
ii. Second way is Smart Card that provide one more level of authentication.

Firstly, both patient and doctors need to register in the system. After registration they can login to the system. Both the users need to securely manage their two type of login credentials (Username password and Smart card). If any one of these two is not available with the user then he/she will not be able to login to smart healthcare system.

The stepwise process is given below:

Step 1: Register user and their details.
Step 2: Activate user account.

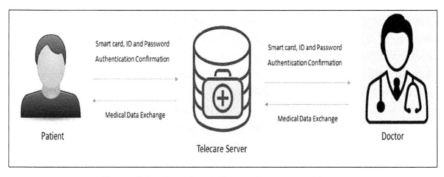

Figure 7.3   Security challenges in smart healthcare

Step 3: Create login module, to log into the system.

Step 4: If patient consults doctor, generate EHR including medical information of patient (diagnosis, treatment, medication).

Step 5: If patient approves the record, store it in cloud database in encrypted form.

Step 6: Store hash of this data in smart contract deployed on private Blockchain.

Step 7: Apply access control.

In Figure 7.4, we have shown the use case diagram where we have presented the detailed processes executed by the system for different users.

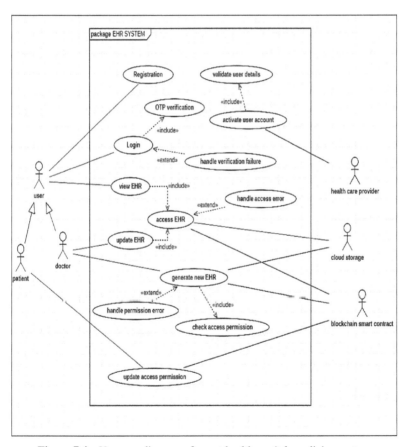

Figure 7.4 Use case diagram of smart healthcare/telemedicine system

7.5 Security Analysis

In this section, we have explained how this can eliminates some security threats with two-way authentication process.

Unauthorised access: During authentication process the user Id is encrypted at user's end before submitting it to the server, hence user ID cannot be obtained by anyone else other than the server.

Resist impersonation attacks: In impersonation attack the adversary impersonates user or server. For launching user impersonation attack, user id is required which is transferred in encrypted form and cannot be obtained without private key of server. For launching server impersonation attack the adversary might intercept message sent to user by server, but at user side timestamp difference is verified first, even if attacker modify the timestamp still some value cannot be obtained and verified without original timestamp sent by server.

Resist man in the middle attack: The server authenticates user by computing values with some mathematical equation. These values can not be calculated without private key of server. The user authenticates server by comparing value of R1, which cannot be done without timestamp, hence these values required for mutual authentication that can be computed by server and user only. Further communication is done with the help of session key, the messages are transferred in encrypted form.

Perfect forward secrecy: If attacker somehow obtains the session key calculated during one session, a new session key is generated for every new session using random numbers (Deffie Hellman).

Resist replay attack: If attacker intercepts the messages shared by user, still replay attack cannot be launched because both the server and user verify the timestamp first.

These are some network attacks and security threats which can be avoided using efficient user authentication scheme

7.6 Conclusion

Telemedical services play important role in the life of patients by providing access to health care services from home. The patient can take advantage of healthcare service from any location and communicate with doctors easily. A patient having medical sensors can share the report with doctors, if patient doesn't have medical sensors, he/she can visit nearby clinic and share the report directly with specialist through telemedical services, hence saving time and expense of travelling. Medical data is very sensitive and confidential; therefore, we need a telemedical system, which ensures the security and access control over medical data. In this chapter, a scheme has been proposed for secure access control of data. As mentioned in informal security analysis, the proposed scheme ensures user anonymity and is secure against impersonation attack, replay attack, insider attack, man in the middle attack and perfect forward secrecy is maintained using random numbers and session key for every new session. The proposed algorithm uses one way hash function, which generates unique output even if there is very small change in input. Concatenation and Bitwise Exclusive OR operations are used which takes very less execution time, hence keeping low time complexity of the algorithm. In performance evaluation, the proposed scheme is compared with other proposed schemes and from informal security analysis it can be seen that this scheme is secure against the listed attacks.

References

[1] A. Saha, et al., (2019). Arijit and Amin, Ruhul and Kunal, Sourav and Vollala, Satyanarayana and Dwivedi, Sanjeev, Review on "Blockchain technology based medical healthcare system with privacy issues". Security and Privacy. 2. 10.1002/spy2.83.

[2] A.T. Wu, B.C. and Chieu, (2003). 'A user friendly remote authentication scheme with smart cards', Computers Security, 22(6), pp. 547–550.

[3] H.M. Sun, (2000), 'An efficient remote use authentication scheme using smart cards', IEEE Transactions on Consumer Electronics, 46(4), pp. 958–961.

[4] S. Salman, et al., (2020). A secure blockchain-based e-health records storage and sharing scheme, Journal of Information Security and Applications 55, 102590.

[5] K. Tanesh, et al., (2019). "AGE: authentication in gadget-free healthcare environments", The Author(s).

[6] M. Zahid, et al., "Authentication and Secure Key Management in E-Health Services: A Robust and Effi-cient Protocol Using Biometrics", Corresponding author: Anwar Ghani.

[7] C. Preeti, et al., (2019). "Cloud-based authenticated protocol for healthcare monitoring system", Springer-Verlag GmbH Germany, part of Springer Nature.

[8] R. Eswaraiah, et al., (2020). "SLT based watermarking system for secure telemedicine" ,Springer Science+Business Media, LLC, part of Springer Nature.

[9] J.C. Patni, et al., (2020). "Pandemic Diagnosis and Analysis using Clinical Decision Support Systems", Journal of Critical Reviews.

[10] S. Gupta, and H.K. Sharma, (2021), "User Anonymity based Secure Authentication Protocol for Telemedical Server Systems", Int. J. of Information and Computer Security.

[11] Shailender et al., (2018). Digital Cancer Diagnosis With Counts of Adenoma and Luminal Cells in Plemorphic Adenoma Immunastained Healthcare System, IJRAR, 5(12).

[12] J.C. Patni, et al., (2020). "Sensors based smart healthcare framework using internet of things (IoT)", International Journal of Scientific and Technology Research 9(2), pp. 1228-1234.

8

Electronic Healthcare Record (EHR) Storage using Blockchain for Smart Healthcare

Abstract

Smart healthcare system require digital records of a patient for diagnosis the disease of the patient. These digital records are called electronic healthcare records (EHR). Since these records are crucial documents for the patient and tempering in these records may harm to the patient health. Efficient storage and security of these records is major responsibility of smart healthcare service provider. In this chapter we have described the use of blockchain technology for efficient and secure storage of these EHR.

Keywords: electronic health records (EHR), blockchain, smart healthcare, telemedicine.

8.1 Introduction

As we know that according to a comprehensive study by Market and Market, the global healthcare market is estimated to reach more than USD 829 billion by the year 2026 from USD 319 billion in 2021 at a compound annual growth rate of 21% during the period under consideration. Medical data is very crucial and sensitive. It is a challenging task to manage patient data on paper. To make patient records available whenever required it is important to maintain digital records and that is why EHRs are maintained. It eliminates the need of searching for previous medical documents while visiting doctor but with this comes a challenge of ensuring the security of this data. Medical data is very sensitive and if changed or tampered by an attacker, may lead to wrong treatment, so this data should not be accessible to everyone, some access control rules need to be applied [1, 2].

8.2 Framework of EHR Storage and Need of Blockchain

Health records of the patients will be stored electronically called electronic health records. The EHR structures manage issues in regards to data security, dependability, and the leaders. EHRs contain fundamental and extraordinarily private information for assurance and therapy in clinical consideration. This data is a significant wellspring of clinical consideration information. An EHR is a plan in the automated arrangement of a patient's prosperity data that is made of all through the patient's life and is usually spread among different centres, offices, and prosperity providers. EHR data will be encrypted using the blockchain and can be shared with the hospitals.

Medical History is one of the most important part of patient's diagnostics. It helps the doctors to give best advice to the patients. All the prescriptions can be stored digitally without the worry of it being lost.

Some keys goals in the implementation of secure blockchain based EHR system are:

- Privacy
- Auditability
- Authenticity
- Anonymity
- Accountability

This data will be highly secured over a decentralized network and cannot be accessed by anyone without the key to access the data [3, 5]. It is a way smarter move to make good healthcare more accessible to the people. This data can be shared with any hospital in the world and the hospital can access it using the key.

In following figure (Figure 8.1), the usages and users of EHR has been presented.

A blockchain is essentially a high-level record of exchange that is duplicated and appropriated across the entire association of hub structures on the blockchain [6, 7]. As blockchain is in effect broadly utilized in various areas, medical care is additionally an essential area where blockchain can be utilized for enhancements reasons. With the popularity of blockchain, there are many applications in the health care sector like the electronic health care system, storage of EHR using blockchain.

Healthcare industry is rapidly and constantly evolving, growing and changing. It is because of the advancement in the technology every day. Governments and even Medical Scientist and trying their best to adapt better ways to provide the finest healthcare services to the people. Implementing

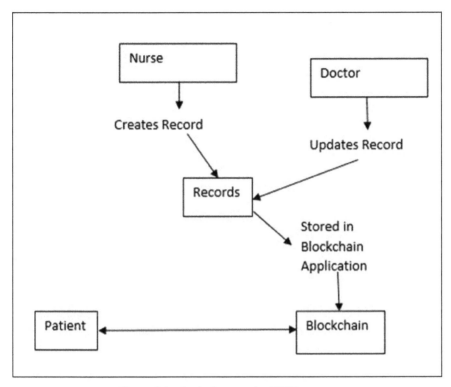

Figure 8.1 Basic framework of EHR storage

new changes with the advancement in technology will allow the doctors and hospitals to diagnose and treat people efficiently and quickly. With the rise in technologies such as AI, ML and blockchain, the occurring of negligence in performing surgeries and accurately diagnosing the patients has reduced to a great extent [8, 9]. New technology called blockchain is in talks nowadays. It is a great technology, which is being used to secure the data and information available without the worry of it being hacked or leaked. It is a digital storage and it is almost impossible to tamper with the data present in blockchain. Blockchain can also be used to electronically storage all the data related to patient's medical history and reports.

Blockchain can lead the way for the smart EHR storage. Even our government can use blockchain technology in keeping the records of patients digitally safe under universal healthcare mission in India [10]. In this mission, government has an intention to keep the medical data, of every single citizen of India, online. Thus, blockchain can pave the way for its security. Blockchain uses a technology named as asymmetric cryptography to keep transactions between users safe digitally. In this system, there will be a personal and a private key for every user. Personal key can be used by medical staff to get some general medical condition of the patient and private key is for the information which the patient wants to keep private and can be seen only with her permission (by using that private key). This can be the idea behind the work of public and private key in blockchain technology. There is a specific technology known as ledger technology, which will be used for the secure digital transfer of the patient's medical records. Even we can store the patient's medical bills, their DNA, blood samples and much more, just in case for the emergency. Many countries are using the blockchain technology in their medical sector. Estonia is one the front forerunners to adopt blockchain technology in 2012 to make their medical sector digitally fast and secure. Currently their entire medical billing, 99% medical prescription and approximately 95% of health data is maintained through Blockchain. The question will rise, when people ask, "Who can see a person's record". The answer is anyone with a public or private key of that transaction can see the detailed information. However, for security reasons, only one person at a time can see the person's medical details and the entities, which are involved in that transaction, can see the details. For example, hospital database system and the patient can be able to access the particular patient's medical history or any medical details. EHR consists of many types of records. This specifically consists of gender, age, health history, vaccination status, allergies status, medicine bills, free government healthcare usage, lab results and so on. Hence, I believe that blockchain has an ability to make our medical transaction digitally faster and safer.

Use of blockchain for medical data may provide enhanced privacy and access control. Patient can have full control over access of their medical records. For security of data in cloud database, EHRs can be stored in encrypted form. The term "Blockchain" is comprised of two words "block" and "chain". We can characterize blockchain as a chain of squares containing data. Blockchain timestamps advanced archives so it is difficult to alter them or change their date. The essential objective of blockchain is to tackle the issue of keeping two fold records without requiring a focal worker. Blockchain is a plan of recording information with the end goal that makes it problematic or hard to change, hack, or cheat the system. A blockchain is fundamentally a high-level record of exchange that is replicated and appropriated across the entire association of hub structures on the blockchain. As blockchain is widely utilized in various areas, medical care is additionally a fundamental area where blockchain can be utilized for enhancements reasons. The EHR structures manage issues in regards to data security, dependability, and the chiefs. EHRs contain fundamental and particularly private information for assurance and therapy in clinical consideration. This information is a significant wellspring of clinical consideration knowledge [11]. An EHR is a plan in electronic design of a patient's prosperity data that is made and stayed aware of all through the patient's life and is conventionally taken care of by and spread among different centres, offices, and prosperity providers. The basic building blocks of blockchain are shown in following figure (Figure 8.2).

We examine how the blockchain innovation can be utilized to change the EHR frameworks and could be an answer of these issues. So basically, EHR systems face a lot of problems. Recent times have shown that the mobility of patients, as well as specialization of health care services, has increased rapidly. To facilitate everything in an efficient way, it is necessary that every hospital and health care centre have access to the patient's medical history so that doctors can take prompt clinical decisions for better diagnosis and treatment. However, EHRs are highly sensitive, and conventional ways such as fax or mail are still being used to share them due to security purposes, which causes delays in patient's care. To overcome this scenario, a system needs to be developed which can share, manage and aggregate data in a secure and trustable manner. For this, we propose a blockchain-based system for EHR data sharing and integration where patients can control the access policy.

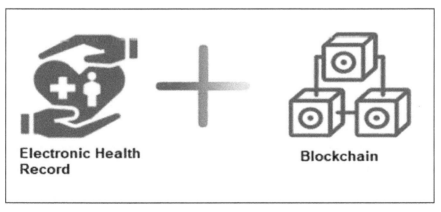

Figure 8.2   Integration of HER and blockchain

Blockchain is a shared, immutable ledger that facilitates the process of recording transactions and tracking assets in a business network. The transactions are stored in the form of blocks, in chronological order, and with an appropriate timestamp. On the content of these blocks, a hash function is applied to form a unique block identifier and eventually forming a chain. Due to hashing property, it can be easily verified if the block was modified and the location of the block can be ensured. The blockchain is replicated and distributed among every participant. Using this decentralized approach, the immutability of the ledger is guaranteed.

A web application can be used by doctors as well as patients to initiate EHR sharing transactions [2]. Every hospital and health care will provide a blockchain node to form a blockchain network and the node integrated with its own EHR system. The shared EHR data will be encrypted and stored in a cloud-based storage. It takes a hybrid management approach, therefore achieving scalability and metadata, on sharing, will be stored in the chain. The patient will be having full control over the data since they will be able to initiate a record-sharing request and define sharing permissions. Digital signature and asymmetric encryption will make this data sharing system very much secure [12, 13].

The successful implementation of the model will result in decrement of data replication and data loss. Also, the accountability on only a single source will be eliminated along with providing high privacy of the data.

8.3 Challenges

It is difficult to manage medical records on paper because the amount of data is huge and it becomes difficult to manage and share this data, hence medical records of patients are stored in the form of EHRs digitally. But with this, comes the challenge of keeping this data secure and at the same time accessible when required. Blockchain technology can be used to apply access control over medical data. Blockchain technology is helpful when it comes to security and keeping authentic log of each and every change in data but we cannot store large amount of data on blockchain because it would consume extra unnecessary space [14]. In such a situation we can store EHRs off chain on cloud database in encrypted form and store other variables, which will be used to decrypt this data and ensure that it is not tampered, in chain with the help of smart contract [15]. Every time any changes are made to medical records, new transaction is generated and hence we have log of every change, also, these transactions are mined i.e., consensus is reached on every node before validating them, hence reducing chances of unauthorized modification of data.

8.4 Conclusion

Nowadays, everyone is moving towards the digital world for doing everything, just like that there are ways by which the health and wellness industry can also use some ways to share the data in a secure manner using the technology known as the blockchain. Blockchain is new to the industry but technology enthusiasts are keeping themselves updated and trying every possible way to implement blockchain; as with help of it; it is impossible to hack the system so keeping it the most secure way. EHRs contain very sensitive and private information related to the treatment in healthcare. Whether to use permission less or permissioned blockchain, it can be decided on the basis of how you would like to proceed with the research. In a permission less blockchain, anyone can easily join the network as anonymous, which will lead to some security issues. A system should provide better security and privacy to the EHR. In permissioned blockchain, their users must have access to the system before joining the network, which leads to security and maintaining privacy in the dataset. A very vital part of blockchain is consensus mechanism, which is used for verifying transactions. If any transaction reaches a particular count of users then that transaction will be added to the consensus.

References

[1] A. Saha, et al., (2019). Review on "Blockchain technology based medical healthcare system with privacy issues". Security and Privacy. 2. 10.1002/spy2.83.

[2] N Kamdar, et al., (2020). Telemedicine: A Digital Interface for Perioperative Anesthetic Care, Anesthesia and Analgesia, 130(2), p. 272–275, doi: 10.1213/ANE.0000000000004513.

[3] S.T. Wu, and B.C. Chieu, (2003). 'A user friendly remote authentication scheme with smart cards', Computers Security, 22(6), pp. 547–550.

[4] H.M. Sun, (2000), 'An efficient remote use authentication scheme using smart cards', IEEE Transactions on Consumer Electronics, 46(4), pp. 958–961.

[5] S. Salman, et al., (2020). A secure blockchain-based e-health records storage and sharing scheme, Journal of Information Security and Applications, 55, 102590.

[6] T. Choudhury, et al., (2021). An algorithmic approach for performance tuning of a relational database system using dynamic SGA parameters. Spat. Inf. Res. https://doi.org/10.1007/s41324-021-00395-5.

[7] S. Gautam, et al., (2019), "A Light and Secure Healthcare Blockchain for IoT Medical Devices", IEEE Canadian Conference of Electrical and Computer Engineering (CCECE).

[8] E. Ahmed, et al., (2019). "Automated Framework for Real-Time Sentiment Analysis", International Conference on Next Generation Computing Technologies (NGCT).

[9] T. Singh, et al., (2017). Detecting hate speech and insults on social commentary using nlp and machine learning. International Journal of Engineering Technology Science and Research 4(12), 279–285.

[10] S. Salman, et al., (2021). "An enhanced scheme for mutual authentication for healthcare services", Digital Communications and Networks.

[11] P. Ahlawat, and Biswas, (2020). "Sensors based smart healthcare framework using internet of things (IoT)", International Journal of Scientific and Technology Research 9(2), pp. 1228-1234.

[12] S. Taneja, and E. Ahmed, (2019). "I-Doctor: An IoT Based Self Patient's Health Monitoring System", International Conference on Innovative Sustainable Computational Technologies, CISCT.

[13] H. Ahmed, et al., (2019). "IoT based smart healthcare for future sustainability", 2019 International Conference on Current and Future trends of IoT.

[14] P.R. Partha, et al., (2020). "Blockchain for IoT-Based Healthcare: Background, Consensus, Platforms, and Use Cases", IEEE Systems Journal.

[15] K. Nirav, and J. Laleh, (2020). "Telemedicine", Anesthesia and Analgesia.

9

Methodologies for Improving the Quality of Service and Safety of Smart Healthcare

Abstract

Smart healthcare and telemedicine systems are network-operated systems. Being a network dependent and network-operated systems, quality of service and safety are some major concern for related to these systems. High availability, High communication Speed, low latency and attack tolerance power is some of the desirable services and safety measurement of these systems. In this chapter, we have described desirable quality of service (QoS) and safety feature in smart healthcare and telemedicine system.

Keywords: quality of service (QoS), security, smart healthcare, telemedicine.

9.1 Introduction

Smart healthcare mainly consists of systems for healthcare like Internet of Things, technology for building medical and health services, a data cloud from medical data [1]. This whole comprises of a very huge process but after that much also, it is a system which lacks security in it and the best solution to this problem is blockchain. It is a new technology but still people out there are keeping themselves updated about blockchain on regular basis as it is the best of keeping a system secure and it becomes almost impossible thing to hack a system after applying blockchain. A smart healthcare system using blockchain can be achieved in three layers transaction layer, information layer and stakeholder layer [2]. Telemedicine services play important role in the life of patients by providing access to healthcare services from home. The patient can take advantage of healthcare service from any location and communicate with doctors easily. A patient having medical sensors can share the report with doctors, if patient doesn't have medical sensors, he/she can visit nearby clinic and share the report directly with specialist through telemedical services, hence saving time and expense of travelling [3]. Medical data is very sensitive and confidential; therefore, we need a telemedical system, which ensures the security and access control over medical data. In following figure (Figure 9.1), various technologies need for providing quality of services to smart healthcare is shown.

We need to ensure user anonymity and security against impersonation attack, replay attack, insider attack, man in the middle attack and perfect forward secrecy must be maintained using random numbers and session key for every new session. We can use one way hash function, which generates unique output even if there is very small change in input. Concatenation and Bitwise Exclusive OR operations can be used which takes very less execution time, hence keeping low time complexity of the algorithm [4].

The idea of savvy medical services has progressively acquired consideration as data innovation advances. Savvy medical services fuses another age of data innovations, like the web of things, enormous information,

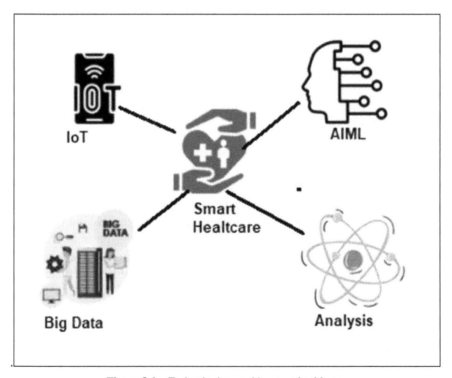

Figure 9.1 Technologies used in smart healthcare

distributed computing, and man-made consciousness, to totally change the current clinical framework, making it more effective, agreeable, and patient-centred. Smart medical care is based on the establishment of data advancements like the Internet of Things, versatile Internet, distributed computing, large information, 5G, microelectronics, and man-made brainpower, just as present day biotechnology [6, 7]. In all aspects of brilliant medical care, these advances are often utilized. Patients can use wearable gadgets to consistently follow their wellbeing and look for clinical help. Patient encounters can be improved by utilizing versatile clinical stages. From the standpoint of scientific research institutes, techniques such as machine learning can be used instead of manual drug screening, and big data can be used to discover suitable participants.

9.2 QoS and Safety Requirement in Smart Healthcare

IoT technologies can be used to increase affected person care. Using IoT-powered far flung care answers that generate and transmit affected person information in real-time in order that the clinical specialists can remotely monitor their patients and hold track of their important symptoms and reply fast in occasion of an emergency, which could finally bring about enhancing fitness results and consequently lowering the affected person's visits to the hospital. Based on one's personal fitness needs, IoT primarily based totally smart healthcare structures can help with conveying custom designed scientific care. IoT-enabled gadgets, for example, smartwatches and bands use sensors to get factors with the aid of using factor and actual time expertise on patients' healthcare, and the facts can be given to medical professionals for additional examination, therefore upgrading the pleasant of medical services [8]. Medical care officers can use the records collected by IoT devices and sensors to differentiate voids in assist deployment and preventing patients' readmission to clinical facilities, alongside loosening the burden of readmissions at the smart healthcare systems [9].

Methodologies that will help in improving the quality of service and safety of smart healthcare

- The rise of artificial intelligence and machine learning will also play a vital role as it can help to deduce insights from patient's data from making various ML models.
- Data security and privacy is also very much required in healthcare systems and here blockchain architecture can help as it will help in securing the databases.

Drug resistance is one of the toughest issues for clinical technology today. IoT may be applied for closely watching a patient's response to a particular remedy and distinguish symptoms and symptoms displaying the presence of medicine resistance. Technologies like ingestible sensors may be used to struggle the problem as they facilitate the gathering of real-time statistics on individual patients. We also can use gadgets consisting of smartphones to boom cognizance of crucial troubles consisting of vaccinations amongst under-vaccinated communities. According to the WHO, vaccine hesitancy, a delay in popularity, or refusal of taking vaccines notwithstanding its availability is one of the pinnacle ten threats to worldwide health [10]. With regards to the smart healthcare industry, IoT has opened a pile of potential outcomes to further develop availability, security and affordability of administrations through better analytic techniques and procedures. The arising technology holds the ability to address different difficulties the smart healthcare industry is confronting today, from rising medical care costs and progressively increasing population and define the manner in which medical services are delivered.

- Utilizing IoT-controlled arrangements and solutions that produce and convey patient information progressively, clinical experts can distantly screen patients and monitor their imperative signs and react rapidly in occasion of a crisis, therefore further developing well being results and lessening the quantity of medical clinic visits of patients.
- Based on an one's own health needs, IoT based smart healthcare systems can assist with conveying customized medical care. IoT enabled gadgets, for example, smart watches and bands use sensors to get point by point and real time knowledge on patient healthcare, and the data can be given to clinical experts for additional examination, consequently upgrading the quality of medical services.
- Medical care officials can use the information gathered by IoT gadgets and sensors to distinguish voids in help deployment and preventing patient readmission to medical facilities, along with loosening the weight of readmissions on the smart healthcare systems.
- Drug resistance is one of the hardest problems for medical science today. IoT can be utilized for intently observing a patient's reaction to a specific treatment and distinguish signs showing the presence of medication resistance. Technologies like ingestible sensors can be used to battle the issue as they facilitate the collection of real-time data on individual patients.

9.3 Methods for Improving QoS

With the advancement and development of technology, the healthcare industry is also making full use of the technologies to change the ideologies of the
traditional healthcare system to make it smarter and people friendly. Everyone
in the world wants to have the best health facilities. The services required by
quality system has been shown in following figure (Figure 9.2).

The speed and accuracy of diagnosis is improving at a faster rate which
is really important for the global citizens. Smart healthcare is made up
of different technologies such as IoT, AI, ML, big data, cloud computing,
blockchain and many more. For example- AI is being used for fast and
accurate detection of diseases. Surgical robots with computer vision are
being developed with perform surgeries with less human intervention and less
negligence. Electronic healthcare record storage is being used to digitally and
securely store all the data related to the patient.

A company called Neuralink is working with AI scientists and medical
scientists to treat neurological problems. The person who has lost the ability
to talk, can communicate with the help of AI. Wearables and other Smart
personal Devices are equipped with heart rate monitors and oxygen monitors
to keep track of everything. The data collected by these devices can be
used to analyses the data and predict whether the person is having any

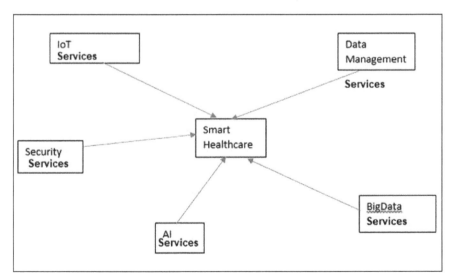

Figure 9.2   Services required in smart healthcare

underlying disease. Various methodologies can be adapted to improve the quality of smart healthcare. Data collected by various IoT and AI devices should be analysed. Data collected is personal to the patients so, it should be secured using high encryption techniques. Self-learning algorithms should be implemented in the devices so that the device can analyze the data itself and suggest treatment options on its own.

The health care industry can be redefined with the integration of IOT. Some ways are:

- IoT technologies can be used to extend patient care beyond the four walls of a hospital. Using IoT-powered for remote care solutions that generate and transmit patient data in real-time so that the medical professionals can remotely monitor their patients and keep track of their vital signs and respond quickly in event of an emergency, which would subsequently result in improving health outcomes and hence reducing the patient's visits to the hospital.
- Based on an individual's unique health needs, IoT devices can deliver personalized healthcare. These devices include smartwatches, smart bands that have sensors to get detailed and real-time insights on a patients' health, and this information can be passed on to a medical professional for further analysis, which would subsequently enhancing the quality of healthcare. IoT devices can be act as a savior for at-risk elderly individuals and for the disabled who need full-time care and attention.
- If we were to collect data using devices we would have a large amount of data that can be used for analysis like for the prevalence of the disease, general health and nutrition status and can even help in identifying the cause for a disease.
- We can also use devices such as smart-phones to increase awareness on important issues such as vaccinations among under-vaccinated communities. According to the WHO, vaccine hesitancy, a delay in acceptance or refusal of taking vaccines despite its availability is one of the top 10 threats to global health.

The idea of keen medical care has step by step acquired consideration as data innovation advances. Keen medical care consolidates another age of data innovations, like the web of things (IoT), large information, distributed computing, and computerized reasoning, to totally change the current clinical framework, making it more productive, agreeable, and patient-focused. Brilliant medical care is based on the establishment of data advances like

the Internet of Things, portable Internet, distributed computing, huge information, 5G, microelectronics, and man-made brainpower, just as current
biotechnology. In all aspects of keen medical care, these advances are much
of the time utilized. Patients can use wearable gadgets to consistently follow
their wellbeing and look for clinical assistance. Patient encounters can be
improved by utilizing versatile clinical stages. From the viewpoint of logical
exploration foundations, methods, for example, AI can be utilized rather
than manual medication screening, and large information can be utilized to
find reasonable members. Smart healthcare actually consists of systems for
healthcare like the Internet of Things, technology for building medical and
health services, a data cloud for medical data. There is a huge process in
Smart healthcare but with a drawback of security in it. A solution for that
is blockchain. Blockchain is new to industry but technology enthusiasts are
keeping themselves updated and trying every possible way to implement
blockchain as with help of it, it is impossible to hack the system, so keeping
it the most secure way.

There are many methodologies which can be used for the improvement of
quality of service in healthcare sector and to secure the medical transactions
in case of their digital being.

Factors that can affect the improvement of quality of service are:

- Start our implementation of improving quality service with a small–
  level demonstration. The reason behind this is the acceptance among its
  various stake holders must be important so as to increase the satisfaction
  of all stakeholders in this case.
- We have to prioritize our basic implementation factors which are to
  encourage communication, engagement and participation of all stakeholders affected by the quality improvement process.
- We have to keep in mind that Quality improvement is an iterative
  process that means results will be reflected in an iterative and continuous
  manner.

One of the best methodologies for improving the quality of service is to
focus on microsystems. In health context, microsystem consists of various
stakeholders such as: a core team of health professionals, Staff who worked
daily to provide basic medical facilities to general public, A work area with
the same clinical and business aims, the shared information environment,
linked processes and shared performance outcomes. Eg. of such team can
be a group of lab technician, primary care providers (nurses etc) and staff of
call centre.

Understanding and implementing improvement cycle consists of different steps:

- Plan strategy: This includes us to prepare for change, establish goals, create teams and much more.
- Develop, test strategy: Select measures to monitor progress, conduct small test of changes, adapt changes to organizational context, identify and deal with barriers and much more.
- Monitor strategy: Implement changes and hold the gains, and evaluate progress against criteria.
- Reassess and respond: Assess with data that what worked and what didn't ans then respond accordingly.

This cycle is known as PDSA cycle, which stands for plan-do-study-act. The PDSA cycle involves all stakeholders in assessing problems (related to its quality) and suggesting (changes required) and testing (those changes) potential solutions. This bottom-up approach increases the chances (probability) that staff will embrace the changes, a key requirement for successful quality improvement.

9.4 Conclusion

Smart Healthcare is one the revolutionary framework that will change the future and will totally change the healthcare system as we have seen it in the past. Integration of advanced IT technologies like AI, IoT, blockchain, image processing, computer vision etc. makes the current traditional healthcare system a smart healthcare system. The complete smart healthcare system is a network-based platform where patient and doctors connect with each other for getting and providing healthcare facility remotely. Providing quality service and safety is major concern for these critical health related systems. In this chapter, we have provided some basic requirements and challenges related to QoS and safety features.

References

[1] P.R. Partha, et al., (2020). "Blockchain for IoT-Based Healthcare: Background, Consensus, Platforms, and Use Cases", IEEE Systems Journal.
[2] K. Nirav, and J. Laleh, (2020). "Telemedicine", Anesthesia and Analgesia, 130(2), p. 272–275, doi: 10.1213/ANE.0000000000004513.

[3] S.T. Wu, and B.C. Chieu, (2003), 'A user friendly remote authentication scheme with smart cards', Computers Security, 22(6), pp. 547–550.

[4] H.M. Sun, (2000). 'An efficient remote use authentication scheme using smart cards', IEEE Transactions on Consumer Electronics, 46(4), pp. 958–961.

[5] S. Salman, et al., (2020). A secure blockchain-based e-health records storage and sharing scheme, Journal of Information Security and Applications, 55, 102590.

[6] T. Choudhury, et al., (2021). An algorithmic approach for performance tuning of a relational database system using dynamic SGA parameters. Spat. Inf. Res. https://doi.org/10.1007/s41324-021-003 95-5.

[7] S. Gautam, et al., (2019). "A Light and Secure Healthcare Blockchain for IoT Medical Devices", 2019 IEEE Canadian Conference of Electrical and Computer Engineering (CCECE).

[8] E. Ahmed, et al., (2019). "Automated Framework for Real-Time Sentiment Analysis", International Conference on Next Generation Computing Technologies (NGCT).

[9] T. Singh, et al., (2017). Detecting hate speech and insults on social commentary using nlp and machine learning. International Journal of Engineering Technology Science and Research 4(12), 279–285.

[10] S. Salman, et al., (2021). "An enhanced scheme for mutual authentication for healthcare services", Digital Communications and Networks.

10

Cloud Commuting Platform for Smart Healthcare and Telemedicine

Abstract

Cloud computing is one of the most prominent technology in information technology that provides virtual computation and storage services at pay on demand basic. The customers need not to purchase physical servers, they just need to generate the demand and same resources will be provided online on rent basis. Cloud computing services are also useful for smart healthcare services. Storage and high availability is one of the major expectation of smart healthcare, which can be fulfil by cloud computing models and services.

In this chapter, we have described the cloud computing service and its significance in smart healthcare to securely store medical records in digital format.

Keywords: robots, smart contract, contact tracing, smart healthcare, telemedicine.

10.1 Introduction

The healthcare industry is one that has historically been behind in the adoption of the bleeding edge of the technological developments. Given the gravity of the work this industry deals with, they have to be significantly more careful with any change made, and have a smaller margin of error [1]. If they unwittingly adopt a technology that was faulty in ways not obvious in the early stages of widespread implement of the same, then it can lead to some disastrous situations. Despite this attitude, the healthcare industry has been one of fastest and willing at adapting the cloud computing technology. From West Monroe Partner's report 35% of healthcare organizations surveyed held more than 50% of the data or infrastructure in the "cloud", or in such a network, as of 2018. Such monumental moves to adopting this technology have not been seen in any other industry. In Figure 10.1, block diagram is shown for data storage and usage through cloud computing model.

Cloud computing, in layman speak, is the ability to store data and use processing power that isn't exactly "yours", i.e. physically a part of your computer. For example, it can allow for barebones computers to execute highly demanding programs or tasks by simply requesting another computer over some sort of network to do it in its stead, deflecting the demands of processing, data and such requirements to the other computer instead. This strips the would-be stringent requirements of the computer hardware of the barebones computer this request originated from, making it easier to perform a larger variety of tasks from "anywhere" given a sufficiently well made network.

This, in general, allows for the smoothening of network structures and the logistics of component acquirement. It also contributes to the management of such a system being easy and intuitive, compared to a system which has no cloud networking components. Another boon of this technology is the data handling methods it has now made possible. Previously, without any cloud,

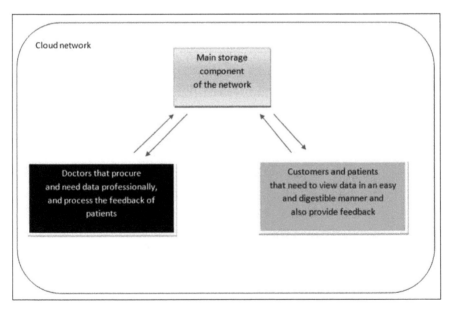

Figure 10.1   Cloud storage for medical data

data that is occasionally would have had to be either stored on site at all times, or rely on wired transmission from another place which has the data. But now with cloud computing, you can make a network with a component that is dedicated to data storage, and connect it to the other members of the network seamlessly. It is now far easier to request the data from the data storage component in the network to receive the required data. With some more optimisation, cloud systems can function as if the data storage component intermediary doesn't even exist, and the data is locally available.

10.2 Cloud Computing

On 16 July 1969 humans created history by launching APOLLO 11, but do you know the computer in the mission had a RAM of 2048 words of memory only and space of 72 kb. The interesting part is our mobile phones and computer systems have far more processing power and memory now but still we feel like if just my ram or processor would have been a little bit better our system could perform little better, but its never enough is it moreover better the processor and memory more is the cost. The cost in case of an individual may be considered as one time investment for say 4-5 years but in the case of companies this cost just keeps on increasing. As they need big server for maintaining their data and to maintain their server they need to invest on a maintenance team that keeps the servers up and running and how can we forget about the Jedi's for the system. That is our team of ethical hacker. What is the only thing standing between a security breech and a well-protected server? Then there is upgradation cost with time technology is evolving and so will the servers and for the company to thrive they will have to upgrade as it's said "IF YOU ARE NOT UPDATED YOU ARE OUTDATED" especially for tech companies they will not prefer to get outdate [2]. So all of this surely cost too much and I can say it's a problem and you know what's the funny thing is about problems, it is that we love to solve them and the solution for this problem presented itself with a beautiful face of cloud computing.

Now cloud computing is a service which is being provided by companies like Microsoft, Amazon, etc. So now you don't have to maintain the servers, you don't have to have ethical hackers to protect your data, you don't have upgrade the servers, you just have to do what you want to do which is just your job and pay for just the services that you require no less no more. Cloud Computing gives you access to huge amount of processing power in even your budget segment system, it gives you flexibility in expansion and elasticity in your business and that too from the comfort of your home or any place

[3]. Cloud computing is not just radically changing the technological fields but even the medical or healthcare system as well. As you see their can be medical records of patient and a ton of files and bundles under which any one could get buried and suffer horribly. But with cloud computing one being used to manage the medical records, patient receipt, medical bills, emr scans. It can be organized and sent in digital format, saves paper and gives a faster access to these records whenever required. Just think if someone suffers from a serious ailment, their medicine receipt and reports lost, and he immediately requires his medicines. Wouldn't it be great if he/she had direct access to that information within his reach in just a few clicks? 12 months ago, the hospitals using cloud computing was just 21% in USA and it increased to 39% in the next 12 months. And its being predicted that in the next 12 months from now their 51% hospitals using this service. Why just talk about USA let us discuss about India. It is predicted that there will be 10% growth every year in this field [4].

Now you must be wondering how this magnificent technology works does. It works differently from front end and back end. Front end gives access to the user who stores their data on the cloud, its basically like interface that makes it easy to interact with the cloud computing software. But we all know that the real movie is made behind the screen by directors, script writers and crew. Similarly the backend is where the data is stored an managed and mad secure for the user, inside a server operations are carried out by following protocols and an amazing software middleware is used to maintain continuous connectivity between the device linked and the cloud. Similarly, the cloud maintains many backup files for user in case of data and security breech to maintain user cloud trust [5].

10.3 Types of Cloud Computing Models

Cloud computing provides multiple models and multiple services to fulfil the computation and storage requirements as per customers need [6].

The various types of cloud computing are:

- **Public cloud:** Public cloud are those type of cloud computing models where it is accessible by all. The resources in this model can be accessed by all peoples who want those resources. This type of model is provided by Google cloud, AWS, Microsoft Azure. When privacy is not a concern than an organization or individual can go for this type of cloud computing model.

- **Private cloud:** This type of cloud computing model is secured and has restricted accessibility. The secret data resources are deployed on this kind of cloud computing model. The charges for this type of cloud computing model is high in comparison to public cloud.
- **Hybrid clouds:** It is built with both public and private clouds. Private clouds can support a hybrid model by supplementing local infrastructure with computing capacity from an external public cloud. An example of hybrid clouds is research compute clouds which as developed by IBM to connect the computing IT resources at eight IT research centres. Hybrid cloud works in the middle of public and private cloud with many compromises in terms of resource sharing.
- **Multiclouds:** Multiclouds are made from more than one cloud services and more than one vendor that is public or private. Multiclouds helps organizations to avail benefits from more than one cloud service provider instead of being dependent on only one service provider.

Various technologies like virtualization, service orientated architecture (SOA), grid computing and utility computing help in making the cloud computing platforms more flexible, reliable and usable [7]. Although cloud computing gives a lot of benefits it also appears to present a risk especially in terms of privacy and security so it is expected that the various cloud service providers understand these issues and look into them seriously. But if we overlook the disadvantages cloud computing has a great role to play in smart healthcare department as well.

10.4 Significance of Cloud Computing in Smart Healthcare

As mentioned in the introduction, data storage is one of primary methods that the industry has put cloud computing to use. It might be slightly un-intuitive, given the name, but just this facet of cloud computing has tremendously helped the healthcare industry. The first thing having such an efficient method of data storage means that everyone involved will be able to retrieve it with ease and elegance. This means that people not familiar with the practice, people like the patients, are easily able to see their own medical reports. Things like mobile applications, more interactive and informative websites are now available due to the ease of transfer and organisation of data as a result of adopting the cloud network in the medical industry. A mobile application can provide seamless and regular updates to patients about their own health, while also reminding of things like when and what medicine to take. This has proven to increase the engagement patients have with their own health, and also has had a positive reaction in general [2]. Another potential benefit to cloud technology is that ease at which a insurance company, or the workplace of the patient can have the information of the medical proceedings in their hands, alleviating the need for the patient to painstakingly draft up a report. All of these are just the beginning of cloud computing, as its applications like EMRs, and other such implementations are being made on a daily basis.

Now onto the "computing" part of cloud computing. Much like other research fields, the medical field has never hurt from more processing power in their computers. More processing power allows for doctors in the field of medical science, and in turn healthcare, to research and investigate more mathematically complex things, and make more accurate simulations. Another use is the ease at which large amounts of data can be processed in the cloud computing network. Having so much hardware ready at hand simplifies these processes and makes it easier for hospitals to diagnose their own treatments and methods by taking in large amounts of feedback from patients and customers.

Finally, we come to the miscellaneous benefits of cloud computing in healthcare. One of the most apparent benefits is the flexibility of this system. Most cloud services are handled by third parties who can handle the large amounts of machinery involved. This allows a very scalable program and no growth pains, unlike when the analogue systems had to be expanded. Other benefits like the maintenance being handled by the provider. The cloud services are generally very affordable as compared to buying your own cloud system as well. All in all, cloud computing can be said to be a unilaterally positive development for the healthcare industry, and a massive plus for humanity in general.

The diagram above is a general and rudimentary display of how information flows in a simple cloud structure. Both doctors and patients need to access similar information, so it is stored in one place, the data storage component of the cloud. The doctors may also internally need to access old data or cross reference data, so to make it easy for all doctors the relevant data is once again stored in the data storage component. The doctors may also need the patient's feedback on treatments and other actions of the hospital, so this data is also stored similarly. The centralization of the data storage makes it much easier to manage, and even in such a simple example in which many nuances are ignored, it is clear how streamlined the data flow is under such an arrangement. Cloud computing is a useful tool, and the fundamental reason for its popularity is that businesses, educational institutions, and other organizations can rent storage, processing capacity, and applications from a cloud service provider. The infrastructure and system maintenance are taken care of by the cloud service provider. For businesses, this means less money spent on computers and fewer employees needed to operate and maintain the system. There are three primary sorts of clouds: public, private and hybrid. Microsoft Azure, Amazon Web Services, Google cloud, and other public clouds are the most popular. The provider owns and manages all of the infrastructure and apps in this scenario. The most significant benefit is that the cloud provider is responsible for all maintenance and management. Private clouds, on the other hand, are restricted to a single firm or organization. The infrastructure for the cloud is often managed by an organization's IT department. The advantages of this strategy for major organizations are that their data is secure, and they can deliver all of the benefits of cloud computing services inside to their many departments, but at a cost. The best value is available with hybrid cloud. The private cloud is used for very sensitive data, whereas the public cloud is used for other services. This gives enterprises more control and security over critical data while yet allowing them to use the public cloud when necessary. Nearly 87% of healthcare firms, for example, choose to employ a hybrid cloud architecture. The private cloud is typically used to host systems that include sensitive patient data, such as imaging and electronic health records (EHR). For storing patients' health files in the cloud, a cloud-based EHR offers a scalable, adaptable, intuitive, and cost-effective solution. We now have the ease of being able to communicate with our doctor on the phone or computer from anywhere in the world, with real-time access to information, anywhere, and compatibility across different systems as a result of this development. Patients' portals, telehealth services, healthcare information exchanges, smart devices, remote monitoring services, and other components of the new healthcare ecosystem are all powered by this. Patients and doctors do not have to be physically present at the same area because of telemedicine and telehealth's virtual care platform. Patients' portals allow patients to access their medical records and contact with their doctors in a safe manner. Cloud computing is a type of internet-based computing that enables virtual access to a variety of applications and services such as storage, servers, and networking. Cloud computing is a whole new virtualization technology for both individuals and businesses that departs from the old software business model. Cloud computing is defined as providing end users with remote dynamic access to services and computer resources via the internet. Medical reports and personal information of the patient is available in the online records and can be easily accessed by the physicians, pharmacies and counsellors without charging any extra cost over the diagnosis fees. The healthcare field has vast, expanse and highly complicated ecosystem including health insurance companies, hospitals, labs, pharmacies/medical stores, patients and other bodies as well hence in order to assure good, effective and fast functionality of this ecosystem it is very essential that data is transmitted quickly and securely between the different entities. Cloud computing provides business models to various hospitals and clinic centres to store the information of the patient and update it with his/her each visit. cloud computing helps in reviewing, exchanging and sharing the images of MRI, CT-scan and X-ray reports in a faster and a highly secured environment. Uninterrupted services can be provided to various health organizations due to high availability of cloud computing services. The various business models of cloud computing benefits the health organizations in many ways. Cloud computing also provides benefits to the patients like improvement in the quality of services and collaborations between various health organizations. Since human life is priceless and precious and the medical facilities across the world are limited therefore the healthcare services provided in cloud offer cost effective concept so as to benefit both the patients and the health organizations. Thus we can see that cloud computing has a big scope in terms of healthcare as well.

10.5 Challenges

Thousands of healthcare organizations have already made the switch to cloud computing services, enhancing their ability to interact with patient data. This transition of medical establishments to cloud computing will undoubtedly grow tremendously. Every major breakthrough in history has been met with a tremendous deal of scepticism and concern at first. Cloud computing is no different.

In Figure 10.2, we have shown the storage and flow of EHR in cloud commuting model.

The issues of cloud computing in healthcare are critical, and they must be addressed efficiently. Furthermore, there are several issues that people are concerned about as a result of this growing innovation, such as:

- **Security concerns:** The security threats are the primary concern for healthcare organizations when it comes to using medical cloud computing. The medical data that hospitals handle, particularly that of patients, is highly sensitive. As a result, the healthcare industry is concerned about data breaches. In the past, there have been attacks against healthcare data

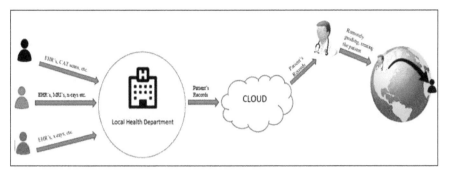

Figure 10.2 EHR storage and flow in cloud storage

servers as well as occasional data leaks. Cloud computing, on the other hand, is improving every day and becoming more capable of withstanding cyber-attacks and system failures. To protect hospitals' medical data, these systems employ security features such as data encryption, security keys, Blockchain integration, and so on.

- **System downtime:** Another major problem with cloud computing is the possibility of system failure. It's similar to when Google or Twitter go down for a few hours because the server's load capacity has been exceeded. Such system downtime incidents are uncommon, but they are a reality that might be a major problem in a 24/7 operating profession like healthcare. We'll receive contingency planning from a professional cloud service provider, such as spare servers to lighten the load and multi-point backups of existing data. The secret to a successful cloud computing service is to build a system that functions even when things go wrong and upgrades when things go right.

- **Limited control:** No one wants their personal data, especially their hospital's medical and patient data, to be controlled by the cloud storage service provider they choose to keep it in. There is no simple solution to this problem. The best thing that can be done right now is for cloud service providers to guarantee that data will not be exploited for personal gain, and there are numerous contracts and agreements in place to ensure that they follow through on their statements.

10.6 Conclusion

Cloud computing is a necessity for high computing and storage systems. Smart healthcare required both feature of cloud computing. For fast response, it needed high computing power and for large storage of medical digital documents, it needed a high capacity storage platform. Due to the features of cloud computing and necessity of Smart healthcare, both can serve the purpose for each other.

References

[1] K. Nirav, and J. Laleh, (2020). "Telemedicine", Anesthesia and Analgesia, 130(2), p. 272–275, doi: 10.1213/ANE.0000000000004513.

[2] P.K.D. Pramanik, et al., Healthcare big data: a comprehensive overview, in: N. Bouchemal.

[3] u, S.T. and B.C. Chieu, (2003). 'A user friendly remote authentication scheme with smart cards', Computers Security, 22(6), pp. 547–550.

[4] H.M. Sun, (2000). 'An efficient remote use authentication scheme using smart cards', IEEE Transactions on Consumer Electronics, 46(4), pp. 958–961.

[5] S. Kumar, et al., (2015). S Dubey, P Gupta, "Auto-selection and management of dynamic SGA parameters in RDBMS", 2nd International Conference on Computing for Sustainable Global Development (INDIACom), 1763–1768.

[6] J.C. Patni et al., "Air Quality Prediction using Artificial Neural Networks", International Conference on Automation, Computational and Technology Management (ICACTM).

[7] T. Choudhury, et al., (2021). An algorithmic approach for performance tuning of a relational database system using dynamic SGA parameters. Spat. Inf. Res. https://doi.org/10.1007/s41324-021-00395-5.

11

Smart Healthcare and Telemedicine Systems: Present and Future Applications

Abstract

Health technology has become an important part for all of us and it plays a major role between patients and doctors. However, as at present it has become a more effective and useful for us. As a technology is increasing every year, the way of operation works is also increasing. The technology standardised physical objects and method for treating and for looking after patient. Technology was intended to improve the quality of healthcare. Medical technology includes medical facilities, information technology and many more. Medical imaging and managing resource imaging have been a long used for medical research, patient reviewing and for treatment analysing.

In this chapter, we have described the current and future application of smart healthcare in diagnosis and prescription of patient health.

Keywords: robots, smart contract, contact tracing, smart healthcare, telemedicine.

11.1 Introduction

Pandemic has introduced technology more vigorously in the healthcare sector and give rise to the resources known as "telehealth" [1]. The person-to-person interaction has reduced to a great extent by the introduction and usage of telehealth. It helps in receiving real time data and information by saving time and patients do not reach to the doctor. It simply require the input from patient and then generate output. Though pandemic has now started coming under control but the rise of telehealth is continuously rising and accelerating [2, 3]. Mostly people are very much comfortable by the usage of telehealth and its

105

wide adoption by the people is the reason for it's success. Hence, it can be said without any doubt that smart healthcare facility has strong future. Now there are many telemedicine apps available in the competitive market. These kind of apps support file transfer, video chats and many more functionalities. "Telehealth" as the word itself tells about the modernisations that has taken place in the healthcare sector and tells us about the capabilities of a machine [4]. It has got very unique way of working because it do the collection of data, processing of data and all these things are done by using computer algorithm [5, 6]. Blockchain is also a latest security trend that need too in integrated in smart healthcare [7, 8].

In following figure (Figure 11.1), a basic model of smart healthcare system is presented.

With increasing population and depleting environmental conditions, many new diseases have paved their way into our lives. But we do not have those many facilities for this large number of population. People from remote places do not have access to these facilities but they do have access to the

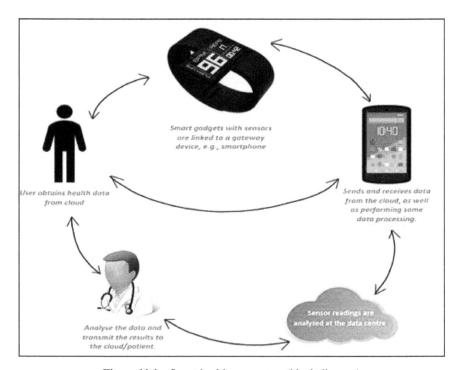

Figure 11.1   Smart healthcare system (block diagram)

technology. And with the help of these technological advancement, everyone is able to access these facilities in any emergency situation or otherwise.

11.2 Current Application of Smart Healthcare

Today is the time of growth. It is also a time of growing pains. New technologies have influenced many parts of our daily life. Today's healthcare system has also recognized the advantages of using information and communication technologies to improve the quality of healthcare, turning traditional into smart healthcare. It ensures that participants get the services they need, help the parties make informed, decisions and facilitates the rational allocation of resources. The present day telemedicine with no special training utilizes existing processing gadgets like cell phones, cameras, wearable biosensors for clinical information, which made it simpler to use. The on-going telemedicine practices for common man lessens travel cost, spares time, decreases clinical expenses and provide simpler smart healthcare specialists access. Information technologies, like IoT, mobile internet, cloud computing, blockchain, big data, 5G, and AI together with modern technologies constitute the keystone of smart healthcare [9–11]. In today's time where we are hopefully getting out of a pandemic implementing tech in healthcare to enhance customer satisfaction, reachability would be a very good step. This implementation requires the cumulative contributions of the government, private organizations. For the current time the implementation of tech in healthcare is not very popular, neither until now it was needed. But, now there is a realization for it. However, the smart healthcare is more oriented towards health management, enhancing patient experience and better drug management rather than disease treatment. In Figure 11.2, the major stakeholders of smart healthcare has shown.

- Healthcare data management is proper.
- Data is immutable.
- Data is provided only by the patients and is transferred only to authorized user and kept away from any unauthorized user.
- Data is safe and secure at various levels.
- Inbuilt protection is present.
- Helps in self-assessment and provides knowledge regarding functionality of our body without any physical test.
- Report made is based on real time input hence there is no chance of wrong report.

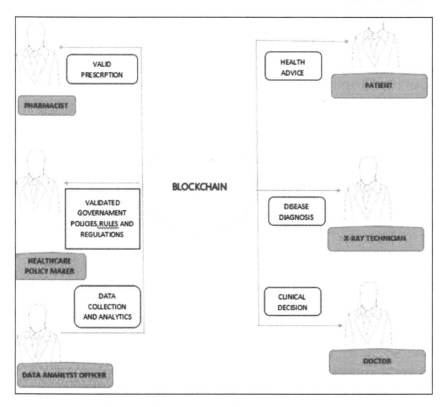

Figure 11.2 Current smart healthcare stakeholders

Current application of Smart Healthcare are given below:

- **Telecare:** Telecare puts an online doctor in the palm of your hand. Before choosing a doctor to consult with, you may use a smartphone and an app to look for doctors online and evaluate their statistics, abilities, credentials, and user ratings. You may consult with the correct doctor through chat or video call after you have discovered them.
- **Health monitoring application:** Individuals and the healthcare sector both benefit from health-tracking applications. They can, for instance, accurately measure blood pressure, heart rate, sleep time, distance travelled, and the number of steps done. They can also help you view your info in an understandable format, store it, perform a statistical study of it, compare your results to standard results, and give you recommendations on how to improve your health.

- **Wearable tech bring in-depth data—driven insights:** As firms develop new ways to measure health and well-being, the wearable technology sector continues to evolve. The Covid-19 epidemic has sped up industrial innovation by giving customized healthcare a tremendous boost. Wearables are now available in a variety of shapes and sizes, including glasses, smartwatches, fitness bands, and fabric, giving customers the freedom to select what best matches their lifestyle and which statistics they wish to follow.
- **Supply chain management:** The most straightforward application of blockchain in health care is to follow the supply chain, such as for conventional pharmaceuticals or drugs containing narcotic compounds that are subject to strict regulation. This is the most realistic scenario, given blockchain has long been utilised in supply chain management. As a drug consumer today, i am unable to determine who manufactured the drug i purchased at the drugstore. You can get a certificate for a medicine when you buy it. But no one thinks it's a good idea for a shady source or merchant to fake it. I can theoretically follow the entire chain if all of the information is maintained in the blockchain from the time of issue until it reaches the retail pharmacy. It is obvious that you can add features to this system in the future, such as drug control in clinics. We can always observe who received the meds, how much of them is needed, and so on. Of course, when it comes to medications used in medical, this is extremely crucial. To do so, a pharmaceutical bar code is printed on the box, which triggers the inclusion of a specific application in the blockchain. Of course, for the user, this system will most likely be a black box in which he would have to put his trust. But it is here that the blockchain second most crucial virtue, immutability, is demonstrated. It is impossible to edit or erase information once it has been stored in it. And this is his major edge over the base, where data can be easily faked, and the fact of the can even be hidden.
- **Data storage for patients:** Day vaccination records, as well as any other medical data, are now recorded in electronic databases on the clinic's side. We, as users, have no systems in place to govern what happens to our medical records. They may theoretically be moved to unidentified groups or simply lost due to a system breakdown. The scenario changes dramatically if the data is transferred to the blockchain. You can create an application that allows the patient to give access to his or her data to accompanying physicians or research groups on a temporary basis. He will be solely accountable for third-party access to his medical card, but

it will be totally apparent to him who uses it for what purposes. Changes. Because such a procedure necessitates the creation of a community, it won't happen tomorrow. However, we can draw parallels with the growth of open apis in the banking industry. This is standard procedure in any bank—you may see inside and examine the transaction. The next important step in spreading blockchain and strengthening the business as a whole may be to create an open API or counterpart of open-banking in healthcare. Data communication across systems is still a major issue to.

Wearables and other home-monitoring technology included in IoT can help doctors track patients' health more effectively. Anyone can monitor patients' adherence to treatment regimens or any immediate need for medical care. IoT assists healthcare professionals in becoming more attentive and proactive with patients. Hence, with no doubt it can be said that "SMART HEALTHCARE FACILITY " has good present and its future is going to be even more stronger. In addition, remote health monitoring help reduce hospital stay time and reduce re-admission. IoT offers healthcare, family, doctors, hospital and insurance business. In IoT wearables and other home-monitoring technology included in IoT can help doctors track patients' health more effectively. Anyone can monitor patients' adherence to treatment regimens or any immediate need for medical care. IoT assists healthcare professionals in being more attentive and proactive with patients.

with the use of recent technologies such as artificial intelligence, IoT surgical robots the diagnosis and treatment of disease have become much more easier and accurate. With the help of robotic surgeries the medical staff are able to perform surgeries which were initially very difficult to perform manually.it also reduce the chances of missed diagnosis and misdiagnosis. Patients are able to get assistance for small diseases from their home itself. Smart healthcare enables the patient to monitor themselves and get immediate feedback and get timely assistance about initial precautions [12, 13]. This will reduce the usage of medical resources, which will help the environment and also reduce health complicacy chances. This is possible with the help of smart watches connected by internet of things technology.

11.3 Future Application of Smart Healthcare

In the future of health, data sharing equitable access, empowered, consumers, behaviour change and scientific breakthrough to collectively transform the existing health system from treatment based care to prevention and wellbeing.

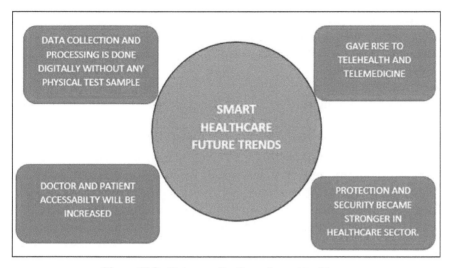

Figure 11.3 Future applications of smart healthcare

Smart hospitals equipped with new technologies, take full advantage of it. These hospitals embed new technologies into their design and operations to improve the customer experiences, as well as outcomes and costs. It connects to a wider healthcare delivery ecosystem, one in which hospitals play as important but less central role. As the technology will increase and be more reliable the number of patients will increase. In future we can use technologies to reach backward areas and give assistance to people sitting at their places which will reduce the travel expenses of the patients and will save their time which would otherwise be spend in visiting hospitals, this would ultimately increase life span of many people due to availability of assistance sitting at one place. This will ultimately maintain quality with low costs [13]. Today we have sphygmomanometer, device to check our glucose level, smart watches etc. Increase in the technology will further help patients to have a home check themselves. Also this will increase the competition thus resulting in more specialised doctor as the patient will go for quality and not availability [14].

In Figure 11.3, we have shown some major future trend of smart healthcare.

The future smart healthcare system will be decentralized and patient-centric:

• Secured and shared personal healthcare record.

- Effective disease prevention and primary care.
- Targeted and better quality acute care.
- Long-term chronic disease management.
- Lower medical cost.
- Anywhere and anytime care delivery.
- Real time data sharing between hospital and patient.

Future application of smart healthcare are given below:

- **Automatic disease diagnosis:** Healthcare organizations already have vast amounts of information about existing medical disorders and available medications. Ai programmes can use this data to produce novel drugs, therapies, and health plans, among other things. Watson for health from IBM and deep mind health from google are two applications that assist healthcare practitioners use cognitive technologies to detect medical disorders based on massive amounts of data. Doctors and pathologists make diagnoses based on the information they have. It goes without saying that a machine learning-enabled system can access far more data than any human person, no matter how experienced or bright.
- **Drug discovery/testing of new drugs:** Because it is a manual procedure, testing novel medications and therapies is slow, resource-intensive, and time-consuming. Artificial intelligence can be employed at any stage of a clinical trial, from finding patients to devising a testing regimen and assessing preliminary results. By automating these simple operations, you may.
- **Surgery aided by a robot:** You are completely incorrect if you believe that artificial intelligence is a new phenomenon in the realm of healthcare. For the past three decades, robots have been utilised to aid during intricate and time-consuming medical procedures. Robots are increasingly assisting in both routine and complex surgeries as a result of the combination of machine learning and artificial intelligence with robotics. They're especially beneficial for repeating repetitive activities with accuracy, reducing the impact of human errors.
- **Remotely nursing assistants:** Many ai healthcare start-ups are working on systems that can provide patients with consultations via voice-based chatting. This can enable rapid access to information on medical issues, obviating the need to see a doctor in many cases. Such solutions can be especially beneficial in the primary health care sector, where doctors and gps are under a lot of strain.

- **Auto-Identification of patients at risk:** As previously stated, hospitals save a large amount of historical patient data. Artificial intelligence can be used to analyse patient health records over time in order to identify those who are at danger. Such technologies can give real-time support with more precision by identifying patterns that put someone at danger owing to poor lifestyle, health problems, or environment. Health insurance companies, who are hesitant to pay for re-admission payments if they occur within 30 days, have prompted the need for patient risk identification. The blame game between healthcare professionals and insurance providers is common, and it's the norm. substantially cut cycle time and, as a result, the cost/effort required.
- **Assistance for administrative workflow:** Artificial intelligence tools such as computer vision, robotic process automation, and remote monitoring are being used by businesses to collect data for a smooth administrative workflow. The information gathered can be fed into analytics to help hospitals make better decisions.
- **Insurer's smart contracts:** Smart contracts, which are mini-programs that run inside the blockchain network, are the third major blockchain application field. They carry out a precise set of actions in response to the triggers they receive at the entrance. Smart stamps have a lot of potential in the healthcare industry. Typically, in the insurance industry, you must go through instances, obtain signatures, and wait a long period for compensation. The majority of the procedure can be automated if the data is recorded on a blockchain that everyone trusts. The smart contract verifies whether all requirements are met and makes a payment as soon as the records are in the chain. If there isn't enough information, he requests that the situation be investigated further. Surprisingly, no public announcements about insurance ventures in medicine have yet been made. However, they are already being talked about in professional circles, so it is only a question of time.
- **Auto-contact tracing:** Fever spikes from their gadgets may be used by smart IoT powered tracking thermometers to monitor disease transmission. This combined data assists in monitoring where an epidemic may arise among the people in their own area.
- **Medical IoT:** Medical IoT is a fast-expanding sector in which wearable gadgets, monitoring, and integrated apps are used to meet healthcare requirements. Medical IoT can offer upgraded versions of classic

medical equipment, such as the smart inhaler—a gadget that synchronizes user consumption with a mobile app—thanks to AI and machine learning technologies.

- **Virtual reality and augmented reality:** Beyond gaming and entertainment, virtual reality (VR) and augmented reality (AR) technology now have a wide range of practical applications. In the medical field, virtual reality aids surgical planning and training, making operations more pleasant for both doctors and patients. There have also been several publications on the effectiveness of VR in the therapy of chronic pain.
- **Blockchain technology:** Blockchain safety and transparency make it ideal for a variety of uses in the healthcare business. Electronic health records (EHRs), remote patient monitoring, the pharmaceutical supply chain, and health insurance claims are just a few examples.

In the future, digital healthcare will be transformative. Many individuals use the internet on a daily basis, and if they can contact with their healthcare providers remotely via their mobile phone or laptop, they are unlikely to visit in person. Prescribed drugs are also delivered to the patient's door. Also we believe that the future is bright, as long as service providers can capitalize on all of the early promise that many technologies are demonstrating. The upward push of m-Health apps from government and wearable devices is presenting get admissions to healthcare experts and sufferers to report their vitals and generate significant insights out of it.

11.4 Conclusion

The notion of smart healthcare has steadily gained prominence as information technology has advanced. When patients and doctors are integrated into a shared platform for intelligent health monitoring by studying daily human behaviours, it may be characterized as smart healthcare. The foundation of smart healthcare solution is the sensors, which records patient's data, stores, process and analyze data, web and mobile apps for patients and caregivers and a gateway to transmit their data. The key challenges in smart healthcare are to reduce operational costs, finish system mistakes, managing disease, boosting up patients experience, better management of drugs and improvement in treatment outcomes. Features of smart healthcare includes PGHD (Patient generated health data), chronic disease management, timely patient monitoring facilities, preventive care, short term care also home-based care which helps the patients and doctors in many ways like time management and transportation costs even and not visiting doctor often in bad state. Smart healthcare has changed all of our lives in many ways.

References

[1] H. M. Hussien, et al., (2019). A Systematic Review for Enabling of Develop a Blockchain Technology in Healthcare Application: Taxonomy, Substantially Analysis, Motivations, Challenges, Recommendations and Future Direction. Journal of Medical Systems, 43(10). doi: 10.1007/s10916-019-1445-8.

[2] K. Nirav, and J. Laleh, (2020). "Telemedicine", Anesthesia and Analgesia, 130(2) p. 272–275, doi: 10.1213/ANE.0000000000004513.

[3] P.K.D. Pramanik, S. Pal, and M. Mukhopadhyay, Healthcare big data: a comprehensive overview, in: N. Bouchemal.

[4] A. Shastri, and R. Biswas "A Framework for Automated Database Tuning Using Dynamic SGA Parameters and Basic Operating System Utilities", Database Systems Journal vol. III, no. 4/2012.

[5] S.T. Wu, and B.C. Chieu, (2003), 'A user friendly remote authentication scheme with smart cards', Computers Security, 22(6), pp. 547–550.

[6] H.M. Sun, (2000). 'An efficient remote use authentication scheme using smart cards', IEEE Transactions on Consumer Electronics, 46(4), pp. 958–961.

[7] S. Kumar, S, Dubey, and P Gupta, (2015). "Auto-selection and management of dynamic SGA parameters in RDBMS", 2nd International Conference on Computing for Sustainable Global Development (INDIACom), 1763–1768.

[8] J.C. Patni et al., "Air Quality Prediction using Artificial Neural Networks", International Conference on Automation, Computational and Technology Management (ICACTM).

[9] S. Salman, et al., (2020). A secure blockchain-based e-health records storage and sharing scheme, Journal of Information Security and Applications, 55, 102590.

[10] T. Choudhury, et al. (2021). An algorithmic approach for performance tuning of a relational database system using dynamic SGA parameters. Spat. Inf. Res. https://doi.org/10.1007/s41324-021-00395-5.

[11] S. Gautam, C. Jorge, and D. Shalini, (2019). "A Light and Secure Healthcare Blockchain for IoT Medical Devices", IEEE Canadian Conference of Electrical and Computer Engineering (CCECE).

[12] E. Ahmed, et al., (2019). "Automated Framework for Real-Time Sentiment Analysis", International Conference on Next Generation Computing Technologies (NGCT-2019).

[13] T. Singh, et al., (2017). Detecting hate speech and insults on social commentary using nlp and machine learning. International Journal of Engineering Technology Science and Research 4(12), 279–285.

[14] S Salman, et al., (2021). "An enhanced scheme for mutual authentication for healthcare services", Digital Communications and Networks.

Index

About the Authors

Dr. Hitesh Kumar Sharma is working as an Associate Professor at the School of Computer Science, University of Petroleum and Energy Studies, Dehradun, Uttarakhand, India. He did his Ph.D. in Database Performance Tuning in 2016. He has completed his M.Tech. in 2009. Currently, he is also working in Machine Learning, Deep Learning, Image Processing and IoT with Blockchain. He has authored more than 60 research articles in the journal and conferences of national and international repute. Dr. Sharma has authored three books and numerous book chapters with various international publishers. He is an active Guest Editor/Reviewer of various referred international journals. He has delivered various Keynote/Guest speeches in India and abroad. He got many certifications in DevOps in the last 2 years. He has also published 03 Patents in his academic career in the last few years.

Dr. Anuj Kumar is an Associate Professor of Mathematics at the University of Petroleum and Energy Studies (UPES), Dehradun, India. Before joining UPES, he worked as an Assistant Professor (Mathematics) in The ICFAI University, Dehradun, India. He has obtained his Master's and doctorate degree in Mathematics from G. B. Pant University of Agriculture and Technology, Pantnagar, India. His area of interest is reliability analysis and optimization. He has published many research articles in journals of national and international repute. He is an Associate Editor of the International Journal of Mathematical, Engineering and Management Sciences. He is also a regular reviewer of various reputed journals of Elsevier, IEEE, Springer, Taylor and Francis and Emerald.

Dr. Sangeeta Pant received her doctorate from G. B. Pant University of Agriculture and Technology, Pantnagar, India. Presently, she is working with the Department of Mathematics of the University of Petroleum and Energy Studies, Dehradun, as an Assistant Professor. She has published around 23 research articles in the journals of national/international repute in her area of interest and is instrumental in various other research-related activities like editing/reviewing for various reputed journals and organizing/participating in conferences. Her area of interest is numerical optimization, evolutionary algorithms and nature-inspired algorithms.

Prof. Mangey Ram received a Ph.D. degree major in Mathematics and a minor in Computer Science from G. B. Pant University of Agriculture and Technology, Pantnagar, India. He has been a Faculty Member for around thirteen years and has taught several core courses in pure and applied mathematics at undergraduate, postgraduate and doctorate levels. He is currently the *Research Professor* at Graphic Era (Deemed to be University), Dehradun, India and Visiting Professor at Peter the Great St. Petersburg Polytechnic University, Saint Petersburg, Russia. Before joining the Graphic Era, he was a Deputy Manager (Probationary Officer) with Syndicate Bank for a short period. He is Editor-in-Chief of *International Journal of Mathematical, Engineering and Management Sciences*; *Journal of Reliability and Statistical Studies*; *Journal of Graphic Era University*; Series Editor of six Book Series with *Elsevier, CRC Press-A Taylor and Frances Group, Walter De Gruyter Publisher Germany, River Publisher* and the Guest Editor and Associate Editor with various journals. He has published 250 plus publications (journal articles/books/book chapters/conference articles) in *IEEE, Taylor and Francis, Springer Nature, Elsevier, Emerald, World Scientific* and many other national and international journals and conferences. Also, he has published more than 50 books (authored/edited) with international publishers like *Elsevier, Springer Nature, CRC Press-A Taylor and Frances Group, Walter De Gruyter Publisher Germany, River Publisher*. His fields of research are reliability theory and applied mathematics. Dr. Ram is a Senior Member of the IEEE, Senior Life Member of Operational Research Society of India, Society for Reliability Engineering, Quality and Operations Management in India, Indian Society of Industrial and Applied Mathematics, He has been a member of the organizing committee of a number of international and national conferences, seminars, and workshops. He has been conferred with the '*Young Scientist Award*' by the Uttarakhand State Council for Science and Technology, Dehradun, in 2009. He has been awarded the '*Best Faculty Award*' in 2011; 'Research Excellence Award' in 2015; '*Outstanding Researcher Award*' in 2018 for his significant contribution in academics and research at Graphic Era Deemed to be University, Dehradun, India. Recently, he has been received the '*Excellence in Research of the Year-2021 Award*' by the Honourable Chief Minister of Uttarakhand State, India.